P9-EEL-488

Spatial Ecological–Economic Analysis for Wetland Management

Modelling and Scenario Evaluation of Land Use

Wetlands are very sensitive and valuable ecosystems that are subject to much stress from human activities. The study presented here develops an innovative triple-layer framework for analysis of wetland management. This approach provides support for spatial matching of physical planning, hydrological and ecological processes, and economic activities. The authors describe how integrated modelling at the regional scale can be achieved in practice. Following an introduction to wetlands, theoretical aspects of the contributing disciplines are discussed, as well as various aspects of integrated and spatial modelling. An applied, integrated assessment of spatial wetland management in the Netherlands, namely for the Vecht area between the cities of Amsterdam and Utrecht, is then presented. This assessment has resulted in a set of linked hydrological, ecological and economic models, formulated at the level of grids and polders, and various types of evaluation and rankings of scenarios. The results indicate the value of maintaining spatial detail for as long as possible.

Written to encompass aspects of both the natural and the social sciences pertinent to environment management, the book will satisfy readers from both areas seeking sustainable solutions at a regional scale.

JEROEN VAN DEN BERGH holds the Chair of Environmental Economics at the Free University, Amsterdam, the Netherlands and is also a full professor in the University's Institute for Environmental Studies. He was awarded the 2002 Royal/Shell Prize for his research on integrated modelling for sustainable development.

AAT BARENDREGT is Lecturer in Environmental Sciences in the Faculty of Geography at Utrecht University, the Netherlands. His research focuses on wetland science, including modelling of wetland ecosystems, and analysis of policy and management to realise sustainable conditions in aquatic systems.

ALISON GILBERT is a research ecologist in the Institute for Environmental Studies, the Free University, Amsterdam, the Netherlands. Her interest in multidisciplinary research, in particular ecological–economic integration, has led to the development and application of analytical tools such as indicators, models, resource accounts and evaluation frameworks.

Spatial Ecological–Economic Analysis for Wetland Management

Modelling and Scenario Evaluation of Land Use

JEROEN C. J. M. VAN DEN BERGH
Department of Spatial Economics and
Institute for Environmental Studies,
Free University, Amsterdam,
the Netherlands

AAT BARENDREGT
Department of Environmental Sciences,
University of Utrecht, the Netherlands

ALISON J. GILBERT
Institute for Environmental Studies,
Free University, Amsterdam,
the Netherlands

PUBLISHED BY THE PRESS SYNDICATE OF THE UNIVERSITY OF CAMBRIDGE
The Pitt Building, Trumpington Street, Cambridge, United Kingdom

CAMBRIDGE UNIVERSITY PRESS
The Edinburgh Building, Cambridge CB2 2RU, UK
40 West 20th Street, New York, NY 10011-4211, USA
477 Williamstown Road, Port Melbourne, VIC 3207, Australia
Ruiz de Alarcón 13, 28014 Madrid, Spain
Dock House, The Waterfront, Cape Town 8001, South Africa

http://www.cambridge.org

First published 2004

Printed in the United Kingdom at the University Press, Cambridge

Typeface Swift 9.5/14 pt. *System* LaTeX 2_ε [TB]

A catalogue record for this book is available from the British Library

ISBN 0 521 82230 0 hardback

Contents

List of figures *page* viii
List of tables x
Preface xiii

1 Introduction 1

1.1 Problems, policy and management of wetlands 1
1.2 What are wetlands? 3
1.3 Wetlands in the Netherlands 5
1.4 Wetland research 8
1.5 Scope and objectives 10
1.6 Outline of the book 13

2 Wetlands and science 15

2.1 Introduction 15
2.2 Hydrology and ecology 15
2.3 Environmental and ecological economics 36
2.4 Natural–social science integration 49
2.5 Conclusions 53

3 Integrated modelling and assessment 55

3.1 Introduction 55
3.2 Integrated modelling 57
3.3 Spatial modelling 62
3.4 Integrated modelling and monetary valuation 69
3.5 Performance indicators 75

3.6 Evaluation 80
3.7 Conclusions 81

4 Theoretical framework and method of integrated study 83
4.1 Objective and approach 83
4.2 Three integration levels 85
4.3 Spatial dimension 87
4.4 Relevance for policy and management 88

5 The Vecht area: history, problems and policy 89
5.1 Description of the area 89
5.2 Historical development of the area 90
5.3 Threats to the wetlands 96
5.4 Economic activities 98
5.5 Policy and management in the Vecht area 99
5.6 Conclusions 104

6 Development scenarios for the Vecht area 106
6.1 Introduction 106
6.2 Scenario I: reference 109
6.3 Scenario II: stimulation of agriculture (Agriculture scenario) 109
6.4 Scenario III: stimulation of nature (Nature scenario) 110
6.5 Scenario IV: stimulation of recreation (Recreation scenario) 112
6.6 Conclusions 114
Appendix 6.1 List of polders 114

7 The spatial–ecological model: hydrology and ecology 116
7.1 Introduction 116
7.2 The hydrological quantity model 117
7.3 The hydrological quality model 121
7.4 The ecological model 123
7.5 Modelling results 126

8 The spatial–economic model: agriculture, nature conservation and outdoor recreation 129
8.1 Introduction 129
8.2 Environmental and economic data 131
8.3 Modelling results under the three scenarios 137
8.4 Economic benefits of nature and recreation 143
8.5 Summary of modelling results 146
Appendix 8.1 Calculation of economic and nutrient variables under the Agriculture scenario 148

Appendix 8.2 Outdoor recreation data 150
Appendix 8.3 Economic scenario results on a polder level 152

9 Performance indicators for the evaluation 162

9.1 Introduction 162
9.2 Performance indicators for environmental quality 163
9.3 Performance indicators for spatial equity 177
9.4 Conclusions 181
Appendix 9.1 Full list of species identified 182
Appendix 9.2 Environmental quality and economic welfare indices on a polder level 187

10 Evaluation of the scenarios 194

10.1 Introduction 194
10.2 Point evaluation 194
10.3 Spatial evaluation 199
10.4 Conclusions 204

11 Conclusions: policy and research implications 207

11.1 Purpose 207
11.2 Method of integrated research 207
11.3 Natural science modelling and results 208
11.4 Economic analysis and results 209
11.5 Indicators, spatial evaluation and scenario ranking 211
11.6 Further research 212

References 214
Index 234

Figures

1.1	Specific features of the case study.	*page* 12
2.1	A wetland plain with a large recharge of groundwater from a nearby hill ridge.	18
2.2	A wetland plain with abstraction of drinking water from a nearby hill ridge.	18
2.3	Soil geomorphology of a wetland plain.	23
2.4	The food web in a fen wetland.	26
3.1	Guideline for the development of indicators.	76
4.1	The integrated approach followed in the Vecht area case study.	84
5.1	The river Vecht with nature at the shoreline and recreation on the water.	90
5.2	The research area and its surroundings.	91
5.3	Land use in the study area and its immediate surroundings.	93
5.4	Succession in a fen area from open water via aquatic vegetation and reedbeds to alder forests.	94
5.5	Mesotrophic quacking fens with a species-rich vegetation of sedges and herbs.	95
5.6	The characteristic landscape of the Vecht area with turf ponds covered with water lillies.	95
5.7	Agriculture in the Vecht area: pasturelands surrounded by nature.	96
5.8	The cormorant, one of the characteristic species in the region, present in a colony of around 10 000 pairs.	98
6.1	The study area and its polders.	107
6.2	Stimulation in the Agriculture scenario.	110
6.3	Stimulation in the Nature scenario.	111

6.4 Stimulation in the Recreation scenario. 113
7.1 Hydrology of the case study area. 118
7.2 Groundwater flow per grid cell per year. 122
7.3 Calculated chloride concentration in surface water per grid cell under the Agriculture (left) and the Nature (right) scenarios. 123
7.4 Calculated orthophosphate concentration in surface water per grid cell under the Agriculture (left) and the Nature (right) scenarios. 124
7.5 The predicted probability of presence of the aquatic plant species *Utricularia vulgaris* per grid cell under the Agriculture (left) and the Nature (right) scenarios. 125
8.1 Schematic view of the various costs related to the Nature scenario. 133
8.2 Schematic view of the various costs related to the Recreation scenario. 136
8.3 Calculation of economic variables for agriculture. 149
8.4 Calculation of nutrient variables. 149
9.1 The evaluation framework. 163
9.2 Vegetation series typical of the succession in the Vecht wetlands. 166
9.3 Interpretation of 'resilience' for succession-driven ecosystems, as defined by (a) Pimm (1984) and (b) Holling (1973). 168
9.4 Aggregation of model output to performance indicators and an index for environmental quality. 170
9.5 The corridor targeted for wetlands restoration within the Vecht study area. 174
9.6 Generation of a performance indicator for spatial equity. 178
9.7 Distribution of welfare per scenario. 180
10.1 Graphical presentation of the effects table and ranking of scenarios. 195
10.2 Trade-off between net present value and environmental quality. 197
10.3 Trade-off between net present value and spatial equity. 198
10.4 Trade-off between environmental quality and spatial equity. 199
10.5 Spatial effects and point effects tables. 200
10.6 Alternative paths for evaluation of scenarios according to objectives with a spatial character. 201
10.7 Welfare per polder. 202
10.8 Difference map for welfare, comparing the Nature and the Recreation scenarios. 203

Tables

1.1 Glossary of selected terms relating to wetlands *page* 4
1.2 Ramsar classification system for wetland type 6
2.1 Definitions and characteristics of concepts used for the assessment of (disturbed) ecosystems 35
2.2 Perspectives of different concepts of disturbed ecosystems with regard to the three ecosystem characteristics of ecological functioning 36
2.3 Differences in emphasis between environmental and ecological economics 48
3.1 Characterising integrated models 63
3.2 Categorisation of possible activities, benefits and costs for a hypothetical wetland change 74
5.1 Inhabitants and surface area of municipalities in or near the Vecht area (January 1996) 92
6.1 Development scenarios for the Vecht area 108
6.2 The 72 polders forming the study area 115
7.1 Input and output of the three applied models 117
8.1 Input and output of the spatial–economic model 130
8.2 Data on types of agriculture 132
8.3 The capacity and capital costs of phosphate-removal plants 134
8.4 Total benefits in the Vecht area 136
8.5 Economic output per region under the Agricultural scenario (changes relative to the present situation) 138
8.6 Environmental output per region under the Agricultural scenario (changes relative to the present situation) 138
8.7 The economic indicators per region under the Nature scenario 140

8.8 Environmental indicators per region under the Nature scenario 141

8.9 The costs and benefits of converting land from use by agriculture into use by nature–recreation 142

8.10 The environmental quality indicators under the Reference and Recreation scenarios 143

8.11 Financial benefits and the net present value under the Recreation scenario including the environmental quality indicator per region 143

8.12 Net economic value of various recreation activities 145

8.13 The economic benefits of converting agricultural land into nature–recreation areas 146

8.14 The net present value per region under the three scenarios 147

8.15 The surplus of nitrogen and phosphorate per region under the three scenarios 148

8.16 Estimated numbers of short holidays 151

8.17 Estimated numbers of long holidays 151

8.18 The economic and environmental output under the Agricultural scenario 153

8.19 The economic and environmental output under the Nature scenario 156

8.20 The economic and environmental output under the Recreation scenario 158

9.1 Performance indicators and index of environmental quality for the four scenarios 173

9.2 Index of environmental quality for polders inundated in the Recreation scenario, and their adjacent polders 176

9.3 Classification of polders on the basis of welfare per polder in the Reference scenario and the average welfare per polder for each class 180

9.4 Application of weights and subsequent calculation of the performance indicator for spatial equity 181

9.5 Plant species identified and used in the construction of the eutrophic species indicator (E) 183

9.6 Plant species identified and used in the construction of the peat accumulation indicator (P) 184

9.7 Plant species identified and used in the construction of the biodiversity indicator (B) 185

9.8 Plant species identified and used in the construction of the non-resilience indicator (R) 187

9.9 Environmental quality (index, [0,100]) and economic welfare (million €) per polder for the four scenarios 188
9.10 Standardised economic welfare and environmental quality [0,100] per polder for the four scenarios 190
9.11 Welfare per polder for the four scenarios 192
10.1 Performance indicators for the three evaluation objectives per scenario: the effects table for the evaluation 195
10.2 Derivation of a single score per scenario according to path 1 in Fig. 10.6 201
10.3 Welfare per corridor polder for the Nature and Recreation scenarios 205

Preface

In all parts of the world, wetlands are endangered by human activities and development. Areas with wetlands often provide locations for housing and recreation. Consequently, threats to wetlands rapidly lead to the loss of the valuable services they provide to humans. Wetlands have been studied in many disciplines, both in the natural and the social sciences. Integration between disciplines has been tried, though often without much success. This study approaches the analysis of wetlands' development and policy by using integrated ecosystem modelling that builds upon a combination of insights from hydrology, ecology and economics. It devotes particular attention to the spatial dimension, the development of a set of complementary indicators and the aggregation and evaluation of information.

The first part of the book provides a short introduction to the relevant building blocks of the approach, which include discussions of wetlands, the natural sciences, economics, integrated modelling and evaluation. The second part of the book presents a case study in which the integrated modelling approach is applied to a wetlands area in the centre of the Netherlands: the Vecht area.

The case study was part of an EU project entitled *Ecological–Economic Analysis of Wetlands: Functions, Values and Dynamics*, sponsored by the EU's Environment and Climate R&D programme (ECOWET, contract no. ENV4-CT96-0273). This project ran from June 1996 to June 1999 and was coordinated by R. K. Turner (CSERGE, C. J. M. University of East Anglia in the UK) and J. C. J. M. van den Bergh. A short article summarising the case study has been published previously (van den Bergh *et al.*, 2001).

We would like to acknowledge the support of various (ex)colleagues. Marjan van Herwijnen, Peter van Horssen, Patricia Kandelaars and Carolin Lorenz have participated in the research and co-authored chapters in the report that forms the basis of the current book. Ernst Bos and Bas Rabeling provided assistance in data collection. Economist Florian Eppink and ecologist Jan Vermaat read critically through parts of the manuscript. Dita Smit edited the final manuscript, and Patricia Ellman checked the (British) English.

1

Introduction

1.1 Problems, policy and management of wetlands

It is now widely recognised that wetlands provide many important goods and services to human societies. Examples include drinking water, flood mitigation, water quality control, fish products and recreational and residential opportunities. The non-use values that society attributes to wetland species and ecosystems can also be significant (Turner *et al.*, 1998a). Wetland ecosystems are, however, under stress from human activities, in particular changes in land use with concomitant habitat loss and fragmentation, resource extraction, drainage and reclamation, and pollution. Not surprisingly, wetlands are currently receiving considerable attention in environmental science and policy.

Wetlands all over the world are threatened, in spite of various international agreements and national policies to protect them. There are a number of fundamental reasons for this (see also Turner *et al.*, 2000). Market failures exist because of the public good aspects of many wetlands and consequent lack of property rights for certain wetland goods and services. In addition, economic activities such as agriculture, industry and water abstraction trigger externalities for other stakeholders. These stakeholders include direct, indirect and even non-users of wetland goods and services. Next, there is a failure of information and a lack of understanding of the multitude of values associated with wetlands as a result of the complexity and 'invisibility' of spatial relationships between groundwater, surface water and wetland vegetation. A final reason is the frequent failure of policy intervention. There is a notable lack of consistency among policies in different areas, such as economics, agriculture, environment, nature protection and physical planning.

A spatial matching among hydro-ecological processes, economic processes and physical planning is needed to reduce the stress on wetlands and to facilitate the return of their goods and services. Integrated wetland research combining the social and the natural sciences can support such a spatial matching and also helps to solve problems of information and inconsistency among various government policies. This book presents an introduction to such research. The sciences most concerned with the study of wetlands are earth sciences (in particular hydrology), biology (in particular ecology) and social sciences (in particular economics). This book focuses on these core disciplines as well as on integration frameworks and modelling approaches. It also offers a detailed account of an empirical integrated study conducted for a region of the Netherlands. This will illustrate both the potential and the problems of the proposed integrated research.

The issue of 'wetlands', most particularly the restoration of wetland ecosystems, is currently topical in the Netherlands, a nation born of the battle to win land from the water. The Dutch strategy towards its wetlands has historically focused on building dykes and developing systems of ditches, canals and pumps to drain the land. Unfortunately, the mismatch between land and water levels is likely to be aggravated by climate change, which is expected to cause a rise in sea level as well as increased precipitation with more frequent and higher floods in winter (Können, 1999). Meanwhile, drainage leads to land subsidence, particularly in areas with peat soils. Drainage exposes the peat component of soils to the atmosphere and triggers its mineralisation. In some areas, soil levels are subsiding by as much as 1 cm per year. Continuation of drainage and constructing higher dykes could very well contain the effects of climate change, but only at a significant cost. This cost includes not only the cost of raising the dykes and greater effort to drain the land but also the consequences of a flood – not if, but when it occurs – and the disaster it will inevitably trigger (Helmer *et al.*, 1996).

There are various social and economic aspects in the debate on water management in the Netherlands. One of the aims of land drainage is to facilitate agricultural activities. This sector is undergoing change as the result of a number of factors. These include reduced European Community (EC) subsidy, growing environmental restrictions and diseases such as bovine spongiform encephalitis (BSE) and the foot-and-mouth epidemic in 2001. Changes in the agricultural sector could make it possible to break the vicious circle of drainage and subsidence, thus opening up land to alternative uses or agricultural practices. One obvious alternative land use is nature, a scarce resource in the densely populated Netherlands. A flexible approach to water, focusing more on adaptation than on defence, is now being considered. Higher water levels in selected areas would seem to be inevitable, and this could offer a number of opportunities for the restoration of wetlands and their goods and services.

Despite recognising these opportunities, changing land use is not easy. The limited experience to date has shown that local stakeholders, particularly farmers, resist change. There are competing goals for nature and a lack of institutional power in the water management community. The limited sense of urgency in some sections of Dutch society constrains land use change, which already needs a long lead-time for its implementation.

Behind these constraints lie questions regarding the gains and losses associated with wetland restoration, the answers to which depend on our knowledge of how wetland systems function. This book is about wetlands, and particularly about raising water levels to restore wetland ecosystems. However, mostly it is about methods and how to analyse and to trade off possible gains and losses from wetland management.

1.2 What are wetlands?

The English language is filled with many descriptive terms for what is becoming generally known as wetlands. Most of us will have heard of terms such as swamp, marsh, mire or bog. These terms often have local meanings that differ across regions and differ again from scientific and legal definitions. A glossary of selected wetland terms is presented in Table 1.1. The term wetlands is, in part, an attempt to encompass all this diversity in a single term. A precise definition poses challenges (Maltby *et al.*, 1996), with more than 50 definitions of wetlands in the literature (Dugan, 1990). Two definitions are offered here to give an idea of why this term remains difficult to specify.

The first is in Article 1 of the Ramsar Convention on Wetlands (held in Ramsar, Iran: Anon., 1997): '. . . areas of marsh, fen, peatland or water, whether natural or artificial, permanent or temporary, with water that is static or flowing, fresh, brackish or salt including areas of marine water, the depth of which at low tide does not exceed six metres'. In short, wetlands are wet land. Land use and land cover have no bearing on the designation of wetlands under this definition. Indeed, wetland conservation and management are not well served by such a broad definition. For example, in these terms, the western part of the Netherlands would be considered wetlands, with Amsterdam and Rotterdam indistinguishable from remnant wetland ecosystems, agriculture, drained polders and lakes.

Natural scientists, and ecologists in particular, would argue that the Ramsar definition of wetlands ignores the physical and ecological processes triggered by water saturation. An alternative definition addresses these aspects (National Wetland Working Group, 1988): '. . . any land saturated with water long enough

Table 1.1 *Glossary of selected terms relating to wetlands*

Bog. A peat-accumulating wetland that has no significant inflows or outflows and supports acidophilic mosses, particularly *Sphagnum* spp.

Carr. A peat-accumulating wetland that receives some drainage from surrounding mineral soil and usually supports shrubs or woods rather than herbaceous vegetation (see fen)

Fen. A peat-accumulating wetland that receives some drainage from surrounding mineral soil and usually supports marsh-like vegetation (see carr)

Marsh. A frequently or continually inundated wetland characterised by emergent herbaceous vegetation adapted to saturated soil conditions. In European terminology, a marsh has a mineral soil substrate and does not accumulate peat

Mire. Synonymous with any peat-accumulating wetland (European definition)

Moor. Synonymous with any peatland (European definition). A 'highmoor' is a raised bog; a 'lowmoor' is a peatland in a basin or depression that is not elevated above its perimeter

Peat. Incompletely decomposed remains of plant and animal life that has accumulated under extremely wet conditions

Peatland. A generic term for any wetland that accumulates peat

Reedswamp. Marsh dominated by *Phragmites* spp. (common reed)

Swamp. Wetland dominated by trees or shrubs (US definition). In Europe, a forested fen or reedgrass-dominated wetland is often called a swamp (see reedswamp).

Vernal pool. Shallow, intermittently flooded wet meadow, generally dry for most of the summer and autumn

Wet meadow. Grassland with waterlogged soil near the surface but without standing water for most of the year

Sources: Mitsch and Gosselink (1993); Pons (1992).

to promote wetland or aquatic processes as indicated by poorly drained soils, hydrophytic vegetation, and various kinds of biological activity that are adapted to a wet environment'.

This second definition provides a useful starting point for the analysis of wetland management. The need for management often stems from economic use, which may trigger or depend on hydrological changes within a wetland. These changes will, in turn, affect the processes emphasised by this definition, even to the extent of precluding them.

For example, wetland ecosystems in the western part of the Netherlands have been subject to agricultural use for some 1000 years. Agricultural use is dependent on the manipulation of water levels, achieved by an extensive system of ditches, canals, dykes and pumps. Is it useful, therefore, to exclude this agricultural land from consideration as wetland because past and present water management is directly aimed toward constraining saturation? While the current management climate is seriously considering, and experimenting with, higher water tables and the return of agricultural land to nature, water levels in the Netherlands will continue to be managed as sea levels rise and the land subsides.

A wide range of different wetland classifications is in use around the world. However, no single classification could be expected to meet all the needs of different wetland inventories (Anon., 2001). The Ramsar classification of wetland types is presented for illustrative purposes in Table 1.2. The primary factors that it uses to distinguish between different types of wetland are the influence of marine waters, which force the presence of salt-tolerant species or halophytes, and the influence of humans.

The wetland ecosystems of the Vecht floodplain, which is the focus of the case study presented in Chapters 5 to 10, may be designated as non-forested and forested peatlands and correspond to categories U and Xp in Table 1.2. Peatlands differ from other wetlands in their combination of interrelated hydrological, chemical and biotic factors, which results in a decrease in decomposition relative to plant production. Organic matter, or peat, accumulates. Peatlands represent an important terrestrial carbon sink, with an estimated 450 Pg, or 25% of the world's terrestrial carbon, currently stored in them (Gorham, 1991; Woodwell *et al.*, 1998). Non-forested peatlands are also known as fens and bogs; forested peatlands are also called carrs. Fens and carrs are typically mesotrophic or eutrophic, with their hydrology strongly influenced by groundwater. Bogs receive their water only from precipitation and are characterised by low water flows. Bogs are acidic ecosystems dominated by oligotrophic species such as *Sphagnum* mosses.

1.3 Wetlands in the Netherlands

The basis for wetland development in the Netherlands was laid some 10 000 years ago. Mean annual average temperatures less than 11 °C, and regular rainfall of about 760 mm per year, meant that there was a surplus of precipitation over evaporation that exceeded 150 mm. This led to waterlogging. These macroclimatic conditions triggered the development of fens, bogs and carrs. Waterlogging was further facilitated by a geomorphology comprising a flat, deltaic landscape of river terraces and glacial moraines. The development of these ecosystems peaked around 3000 years ago; since when they have declined drastically. This decline has been particularly severe over the last 1000 years and can be attributed to the influence of humans (Pons, 1992). This influence began with catchment clearance and its impact on river discharges more than

Table 1.2 *Ramsar classification system for wetland type*

Marine/coastal	
A	Permanent shallow marine waters
B	Marine subtidal aquatic beds
C	Coral reefs
D	Rocky marine shores
E	Sand, shingle or pebble shores
F	Estuarine waters
G	Intertidal mud, sand or salt flats
H	Intertidal marshes
I	Intertidal forested wetlands
J	Coastal brackish/saline lagoons
K	Coastal freshwater lagoons
Zk(a)	Karst and other subterranean hydrological systems
Inland wetlands	
L	Permanent inland deltas
M	Permanent rivers/streams/creeks
N	Seasonal/intermittent/irregular rivers/streams/creeks
O	Permanent freshwater lakes
P	Seasonal/intermittent freshwater lakes
Q	Permanent saline/brackish/alkaline lakes
R	Seasonal/intermittent saline/brackish/alkaline lakes and flats
Sp	Permanent saline/brackish/alkaline marshes/pools
Ss	Seasonal/intermittent saline/brackish/alkaline marshes/pools
Tp	Permanent freshwater marshes/pools
Ts	Seasonal/intermittent freshwater marshes/pools
U	Non-forested peatlands
Va	Alpine wetlands
Vt	Tundra wetlands
W	Shrub-dominated wetlands
Xf	Freshwater, tree-dominated wetlands
Xp	Forested peatlands
Y	Freshwater springs, oases
Zg	Geothermal wetlands
Zk	Subterranean karst and cave hydrological systems
Man-made wetlands	
1	Aquaculture ponds
2	Ponds
3	Irrigated land
4	Seasonally flooded agricultural land
5	Salt exploitation sites
6	Water storage areas
7	Excavations
8	Wastewater treatment areas
9	Canals and drainage channels, ditches
Zk(c)	Karst and other subterranean hydrological systems

Source: Anon. (1997).

2000 years ago, but has been greatest with human settlement over the last 1000 years. The main human use of wetlands has been for agricultural purposes. Conversion of wetlands to agriculture initially required their drainage and subsequently the continuous management of water levels.

Today, the remnant wetlands of the Netherlands are made up of a number of different types. The western half of the Netherlands is under the direct influence of the sea and rivers. Water tables are close to the surface and the sedimentation of clay is prominent. Hundreds of years ago, these areas were frequently under water as a result of flooding from the sea and from rivers. Today, they are constrained by dykes and drainage systems. Between the coastline with its dunes and the eastern parts of the country there was once a large plain where wet conditions supported the development of marshes, fens, carrs and bogs.

The soil in the eastern half of the Netherlands consists of sandy layers. Runoff is brought to the main rivers via streams and small rivers, another type of wetland. Small lakes can result if water is still or stagnating. These conditions promote the development of bogs, where the permanently high water tables with nutrient-poor and acid conditions stimulate the growth of mosses (*Sphagnum* spp.). Historically, bogs covered large areas of the Netherlands. Only remnants now remain.

The Rhine, Meuse and Scheldt rivers are fine examples of riverine wetlands. Longitudinal and transverse gradients differentiate these wetlands. Where the rivers enter the Netherlands, the riverbeds are sandy, but the clay component increases towards the coast as stream velocity decreases. Close to the sea, there is a change from fresh to tidal-fresh to brackish wetlands. Transverse gradients – from the centre of the permanent river into the floodplain adjacent to the river – are linked to river discharges. River discharges peak at the end of the winter when the floodplain, bounded by winter dykes, is submerged. Conversely, the smaller discharges in summer mean that these riparian areas dry out. This creates wetlands that are periodically flooded and which are rich in nutrients and minerals; that is, they are very fertile. Apart from the present rivers, remnants of former river courses (e.g. oxbow lakes), as well as ponds that formed when rivers broke through their dykes (in Dutch: *wielen*), may also be found.

The Rhine, Meuse and Scheldt form a delta where they meet the shallow North Sea, resulting in an estuary with a wide range of fresh to brackish or saline conditions. To the north of the delta lies a long coastline characterised by sandy beaches backed by dunes. Further north and east lies the Wadden Sea, with its islands, sandbanks and channels. Tidal marshes, sandy banks, islands, gullies and deeper streams create a patchwork of wetland systems in both the delta and the Wadden Sea.

Water management has always been of paramount importance to the Netherlands. With the increased value being placed on nature in recent decades combined with concerns for the future with possibly serious impacts of climate change, the management of its remaining wetlands and wetland restoration are being given much attention and priority on research as well as policy agendas.

1.4 Wetland research

Existing wetland research can be classified in many ways, including by a focus on:

- hydrological versus ecological processes;
- natural versus social science issues;
- social versus economic problems;
- monodisciplinary versus integrated approaches;
- temperate versus tropical wetlands;
- coastal versus freshwater wetlands;
- single versus multiple use;
- providing services to 'nature' (links in ecological networks, habitat for migratory birds) versus services to humans; and
- nature conservation versus ecosystem engineering.

Wetlands are not just valuable and sensitive ecosystems but are also very dynamic and adaptive systems. This means that, even under pristine conditions, wetlands can significantly change over time in terms of type of vegetation, density of vegetation, type of fauna and open versus closed area. Various frameworks have been designed to address the integrated analysis of ecosystems in general and wetlands in particular (e.g. Berkes and Folke, 1998; Turner, 1988; see also Section 2.4). Terms used in this context are 'attributes' or 'characteristics' (biological, chemical and physical), 'structure' (tangible 'elements' such as plants, animals, soil, air and water) and 'processes' (transformation of matter or energy) (see Turner *et al.*, 2000). Processes include the interactions between wetland hydrology, geomorphology, soil and vegetation (Maltby *et al.*, 1996). Given stable interactions, the provision of goods and services (or functions) can be maintained. Economists have added the notion of values to these goods and services. These are usually classified into use and non-use or passive-use values (Barbier, 1994; Gren *et al.*, 1994). Services, goods and values can be attributed to a range of stakeholders (Turner *et al.*, 2000):

- direct extensive users, who harvest wetland goods in a sustainable way;
- direct exploiters, who extract wetland physical resources in ways that possibly damage the wetlands;

- agricultural producers, who affect water levels in wetlands by the drainage of agricultural land and cause the emission of nutrients to water and soils;
- water abstractors, who affect wetland water tables by using scarce water for drinking water and agricultural irrigation;
- communities close to wetlands, who drain wetlands for land use opportunities (housing, industry, infrastructure);
- indirect users, who enjoy wetland services, such as storm abatement, flood mitigation and water purification; and
- nature conservation groups.

In setting up wetland research, a number of considerations are relevant. One has to decide about the appropriate terminology and typology of wetlands, their functions and their values. This is important to avoid confusion between interpretations of terms and concepts by researchers with different disciplinary backgrounds. Terms that have at times generated confusion are, among others, 'threshold', 'stability', 'equilibrium', 'function', 'scale' and 'value' (e.g. Muradian, 2001). In addition, one has to demarcate the range and scale of effects to be analysed and assess possible associated thresholds. This is important to keep the analysis both relevant and feasible. This enables one to adopt a local, regional, national or supranational approach (Gibson *et al.*, 2000). Next, one has to assess the causes and mechanisms of wetland change, degradation and loss. Distinguishing between proximate and ultimate causes, and between immediate causes and historical context, can be helpful in decomposing the complexity of relationships. The choice of research methods, notably integrated modelling, spatial modelling, ecological and economic valuation, performance indicators, and evaluation procedures, is an important subsequent step. It defines the framework of integration of natural and social science concepts and data. This is useful in getting to grips with the complexity and 'invisibility' of spatial relationships between groundwater, surface water, wetland vegetation and economic values (Turner *et al.*, 2000). Finally, aggregation of economic and environmental indicators can be done through a multicriteria evaluation procedure. It has the advantage that no fixed aggregation function is employed to arrive at a single index, but instead the indicator and spatial aggregation – including choice of weights – can be explicitly based on expressed preferences by private or public decision makers.

The design of instruments for local wetlands management and regional, national and international wetland policy is useful both for the development of scenarios to be studied with the resulting integrated model and for the practical implementation of study findings. The latter should take into account market and policy failures, such as wrong incentives (e.g. subsidies), lack of monitoring and control, lack of property rights or simply lack of policies in general (Turner and Jones, 1991). The inclusion of stakeholder interests, behaviour, conflicts and cooperation, as well as distributional issues in the analysis, can increase the social feasibility of any policy recommendations of the research. Distributional issues can be related to social groups as well as spatial zones or regimes. Ultimately, policy outcomes depend on the selected goals of the analysis for wetland policy and management: efficiency, welfare (including the issue of welfare to whom), costs of public regulation, sustainability, nature conservation, multiple use, etc.

These various considerations explain the wide variety of wetland studies in the literature. From a social science perspective, monetary valuation studies (Barbier *et al.*, 1997; and Section 2.3) and integrated modelling (see Ch. 3) are the two most important categories. In some cases, these approaches can serve as substitutes, while in other cases they can be complementary (e.g. Costanza *et al.*, 1989; and Section 3.4). As a third category, one can consider transferability of information and results of one empirical study to another context, region or country. It requires the choice of transparent methods and indicators. Specific methods such as value transfer, which is based upon statistical (meta-) analysis of past studies, allow for cost-effective generation of present studies (see Section 2.3.2).

As explained in Section 1.1, hydrology, ecology and economics are most crucial to the study of wetlands. It was already noted that a spatial matching among physical planning, hydro-ecological processes and economic processes is able to reduce the stress on wetlands. It can be supported by integrated modelling and scenario analysis and evaluation. This combination can provide information about possible future wetland development, subsequent economic (e.g. efficiency) and environmental (e.g. biodiversity) conditions, including their spatial dimensions, as well as suggest priorities within this development. In the 1990s, there have been many wetland studies focusing on ecological questions, hydrological issues or economic monetary valuation (e.g. Brouwer and Spaninks, 1999; Gopal *et al.*, 2000; Mitsch and Gosselink, 1993; van Dijk and Kwaad, 1998), but very few really integrated studies (see Turner *et al.*, 2000). Chapters 2 and 3 will provide more detail on wetland research as well as on integrated modelling and analysis.

1.5 Scope and objectives

This book discusses the integration of information, concepts and models from the social and natural sciences in the context of wetland research. This integration is applied to the analysis and evaluation of alternative paths for regional development in a particular area. These paths are driven by changes in land use and cover and are analysed by integration of spatial hydrological, ecological and economic information. The specific details of the integration include:

- consistency of hydrological, ecological and economic settings at the spatial scenario level;
- linking of hydrological and ecological models (together forming the ecosystem model);
- linking of economic and ecosystem models;
- integration of information in aggregate indicators; and
- integration of biological and economic indicators through point and spatial evaluation.

The region chosen for this application is the Vechtstreek, an area of wetlands in the western Netherlands. It comprises the floodplain of the river Vecht, located in the centre of the Netherlands between the cities of Amsterdam and Utrecht. About half of the area lies in the province of North-Holland and half in the province of Utrecht. The area is approximately 8 km by 20 km and comprises many natural and artificial lakes, reedbeds and marshes, as well as wet meadows. It is predominantly peatland, with 0.5 to 3 m of peat lying above a sandy subsoil. Many characteristic bird and plant species are present. In this area, the value of nature is high, in both national and international contexts: Naardermeer, a lake in the north of the study area, is designated as a Ramsar wetland; the region forms part of the Main Ecological Network of the Netherlands; and the river Vecht belongs to the Blue Axis of the Netherlands (this refers to policy that stimulates the wet ecosystems in spatially linked corridors across the country).

From a hydrological perspective, the whole of the Vechtstreek can be regarded as wetlands since the groundwater table reaches the surface almost everywhere. From an ecological perspective as well, most of the area can be identified as wetlands, since typical wetland vegetation is found both in areas under agricultural use and in natural areas. In terms of the Ramsar classification of wetlands (see Table 1.2), much of the area falls under the category of 'man-made wetlands, category 4': 'seasonally flooded agricultural land'.

Specific features of the case study described in this book derive from a number of factors, as shown in Fig. 1.1. The driving influences within the physical system are the different sources of water: both their quantity and quality. The size of the area, and particularly the heterogeneity of physical characteristics and processes, is such that a valuation study is extremely difficult. Valuation studies seem more suitable for areas that are both smaller and more homogeneous

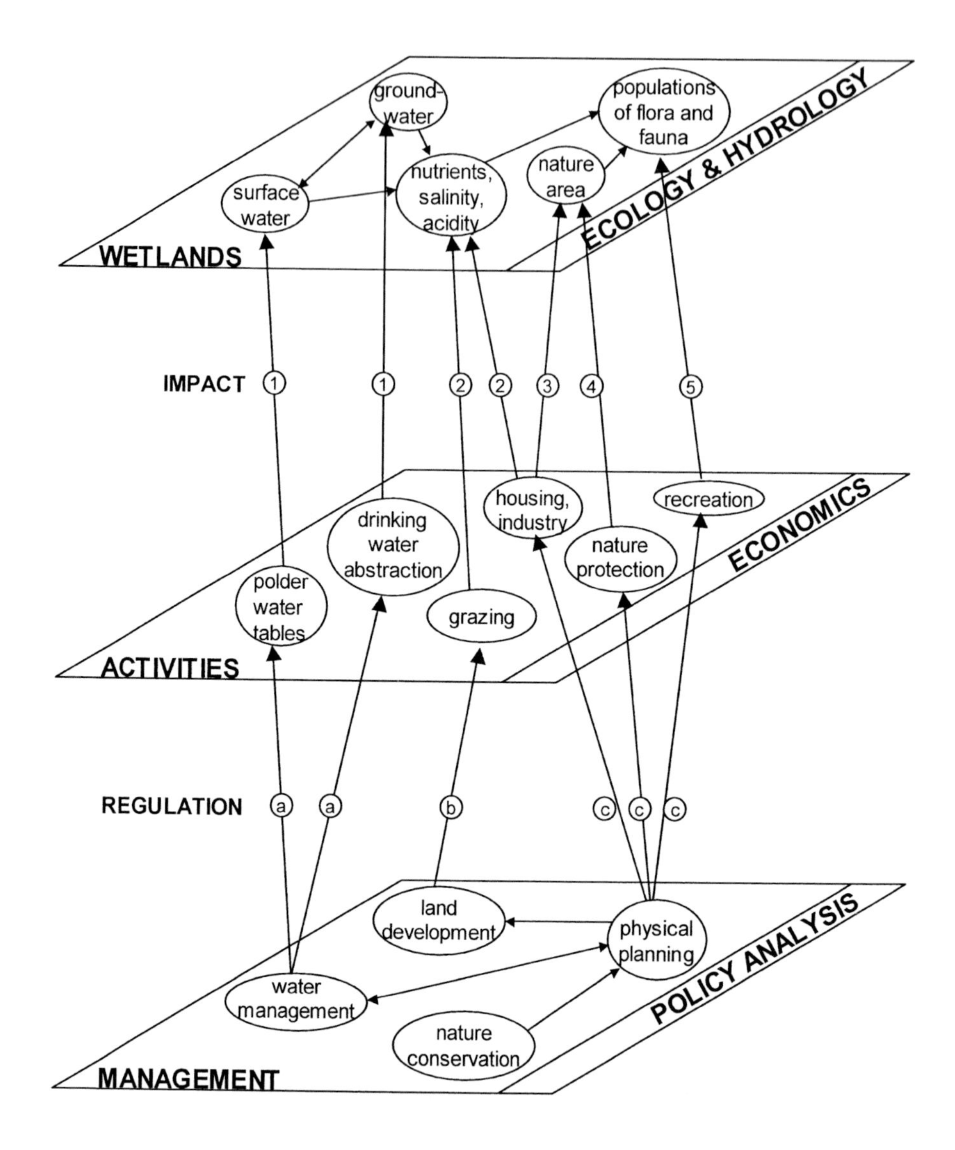

Regulation of economic activities

a Water Boards
b Land Development Commissions
c Municipal, provincial and national government

Impact on physical wetland system

1 Desiccation/dehydration
2 Eutrophication
3 Fragmentation
4 Nature restoration
5 Disturbance

Fig. 1.1 Specific features of the case study.

(e.g. Gren *et al.*, 1994). Land use in the Vecht area is also diverse, including housing, infrastructure, agriculture, recreation and nature conservation. Historically, the region's development focused on agriculture. However, there is now considerable societal support for the restoration of hydrological and ecological gradients. A number of small restoration projects are currently being undertaken by governmental and non-governmental organisations. Finally, many different local, provincial and national governments control the legal and institutional processes that influence the status and development of the area.

The present study aims to develop a method for assessing regional development that is based on spatially disaggregated scenario analysis. This involves the following steps.

1. Construction of a spatial model of the Vecht river basin, representing economic activities, hydrological processes and ecological responses.
2. Formulation of scenarios on the level of grids (square spatial units of 500 m × 500 m) and polders (on average 200 ha), reflecting alternatives for development in which current plans for nature conservation and restoration can be implemented.
3. Development of indicators for economic efficiency and environmental quality.
4. Point and spatial evaluation of the scenarios with respect to these criteria.
5. Identification of constraints on, and opportunities for, a better balance between the region's physical processes and its economic activities.

1.6 Outline of the book

The book consists of 11 chapters. Chapter 2 gives a short introduction to the disciplines that have contributed to the present integrated research study. These include hydrology, ecology and environmental economics. Special attention is given to functions and processes in wetlands, soil conditions and processes, general issues in hydrology, particular features of wetland hydrology, ecological evaluation and concepts of ecological and environmental quality. In addition, the central insights of environmental economics, the distinction between ecological and environmental economic perspectives, and the integration of natural and social science aspects in research are discussed.

Chapter 3 follows with a discussion of tools, focused on integrated modelling and assessment. This covers not only the main literature in this area, relating to global and regional modelling, but also a number of aspects that are of particular relevance to the present study. These include the spatial economic and ecological dimensions of integrated models, the various possible relationships between monetary valuation and integrated modelling, the construction and incorporation of performance indicators, and the choice of concepts and use of methods for the evaluation of output generated by integrated models.

Against the background created by these two relatively long chapters, Chapter 4 presents a concise discussion of the framework of integrated research adopted in the present study. It clarifies the three different levels at which integration occurs, as well as the particularities of the spatial dimension of the study.

Chapter 5 introduces the reader to the Vecht area case study, by providing a brief description of the study area, its history, its problems, its economic features and the range of implemented policies that affect it.

Chapter 6 presents four scenarios for analysis, giving attention to both general and spatial characteristics. In addition to a reference case, these scenarios focus on providing attractive conditions for a particular type of activity: agriculture, nature and outdoor recreation.

The next two chapters present the spatial integrated model, along with disaggregated results. The natural science dimension of the model is explained in Chapter 7. There are three submodels, namely the hydrological quantity, the hydrological quality and the ecological model. The economic dimensions are discussed in Chapter 8. The chapter focuses on the three economic activities – agriculture, nature and recreation – which are also reflected in the choice of scenarios.

The disaggregated results of running the different scenarios through the models are subsequently aggregated along different dimensions in Chapter 9. This involves the construction of environmental and economic performance indexes. The construction of an environmental quality index based on a number of ecosystem performance criteria receives special attention.

Chapter 10 then presents applications of point and spatial evaluation procedures, using calculations based on the available indexes. This starts from the idea that three major criteria or, associated with these, conflicting objectives can be identified: maximisation of net present value, maximisation of environmental quality and maximisation of spatial equity. The results of this exercise are rankings of the social desirability of scenarios.

General conclusions are drawn in Chapter 11.

2

Wetlands and science

2.1 Introduction

This chapter provides a concise introduction to the elements – concepts, theories, methods and models for empirical analysis – contributed by the natural and social sciences to studies of wetlands. The case study discussed in Chapters 5 to 10 of the book will draw upon some of these elements. Section 2.2 presents the natural science perspective, focusing on hydrology and ecology, while Section 2.3 presents the social science perspective, focusing on environmental and ecological economics. Section 2.4 examines approaches to, and problems associated with, integrating the two perspectives.

2.2 Hydrology and ecology

According to the Ramsar definition, a wetland is an area where wet conditions predominate (see Section 1.2). Abiotic – chemical and hydrological – conditions of waters and soil together influence biological processes in a wetland, which in turn affect its environmental and economic functions. This section will provide information about these various connections, illustrated by examples taken from the Netherlands.

2.2.1 *Processes and characteristics of hydrological systems*

Each wetland is part of the hydrological cycle. The latter refers to the process of water evaporating from the surface of the earth and returning through precipitation: rain, snow, etc. The volume of precipitation is the major force in the cycle. Since gravity drives water flows, surface water ends up in rivers and oceans, where it can evaporate again (Chow, 1964; Manning, 1997; Ward and Robinson, 2000).

Water cycles can be represented through balance equations. For the whole earth, precipitation (P) is equal to evaporation (E). On a continental or national scale, precipitation minus evaporation equals run-off (R), the amount capable of entering rivers and flowing to the ocean. For example, the average water balance for the Netherlands is as follows. It receives precipitation of approximately 780 mm/year. Evaporation is estimated at 530 mm/year, implying that run-off equals 250 mm/year. In addition, the Netherlands receives a large continental run-off of approximately 2000 mm/year because of its situation on the delta of two large continental river basins (the Rhine and the Meuse). Note that, since the Netherlands is rather flat, the surplus of water creates run-off problems, which explains why large parts of the country are characterised by wet conditions.

The water balance of an open hydrological system, such as a lake or river catchment, a coastal delta, or a wetland basin, is affected by a number of positive and negative changes in the storage of water (V). As a result, the water table can fall ($V > 0$) or rise ($V < 0$). Water input includes not only precipitation (P) but also flows of discharging groundwater (G) from other areas that enter the system and direct inflows of surface water (S). Negative influences on the water balance are groundwater infiltration of the soil as a recharge (I) and discharge to a river (D). Finally, water can disappear through evaporation (E) from open surface water or through evapo-transpiration via vegetation (T). Hence, the water balance for the system can be written as:

$$P + G + S = V + I + D + E + T.$$

Many studies use this balance to study the hydrological circumstances in disturbed and stationary systems. However, not all variables in the equation can easily be assessed. Precipitation and evaporation can be measured quite simply. Although evapo-transpiration is difficult to measure, estimates for particular areas can be obtained by taking empirical findings in the literature and adapting these by adding information about meteorological circumstances and vegetation in the area under consideration (Koerselman and Beltman, 1988). The flow in surface water is equal to flow velocity times the diameter of the flow channel, which can both be assessed relatively easily at locations with input and output. The only variables that cannot be directly measured are discharge and recharge by groundwater.

Nevertheless, these latter two elements of a water balance can be inferred indirectly through hydrological modelling. This requires the explanation of a number of key concepts. Groundwater can be divided into saturated and unsaturated zones in the soil. The unsaturated zone is the upper part of the soil. From here, water can percolate down to zones with capillary conditions. Infiltrating water finally ends up in the saturated zone, where 100% of the pores are filled with water. Water flow is retarded by soil resistance. This involves a pressure that results from higher levels of groundwater waiting to flow into lower levels.

The flow of groundwater can be described according to Darcy's law:

$$Q = ci,$$

where Q is the flux of water, c is the resistance to flow in the actual soil and i is the gradient of groundwater pressure. The variable c is a function of the characteristics of the soil and depends on pore space in the soil, which differs between clay, peat, sand or gravel. Sand and gravel have the lowest resistance to groundwater, with at least 30% of total volume being pore space. Clay and peat have the highest resistance to groundwater flow, with a pore space of not more than 1% of total volume. The flow of water per metre is measured in days or months in coarse sandy soils and in years or centuries in heavy clay types.

The piezometric head, or groundwater pressure, can be measured by inserting plastic tubes (piezometers) in the groundwater and measuring the water level compared with a fixed elevation. In this way, lines of equal pressure, known as isohypses, can be assessed. Lines perpendicular to these isohypses indicate the gradient of groundwater pressure, as well as the flow direction of the groundwater flow: namely, from locations with high pressure to locations with low pressure. At some locations, groundwater pressure can be very high as a result of aquitards, such as clay layers, which reduce the possibility of flow of groundwater into nearby aquifers.

Modelling the groundwater flow, therefore, involves the measurement of groundwater levels at a large number of locations and the choice of boundary conditions. Once this information is available for a set of soil layers, a three-dimensional pattern of flow can be calculated (Ward and Robinson, 2000). If such a model describes the current groundwater flow accurately, it can be used to predict a number of characteristics of the groundwater system, such as the spatial pattern of recharge or discharge. From the pressures in the first, second and subsequent aquifers, the flow of groundwater can be calculated in terms of direction and flux (Anderson and Woesner, 1992; Hill, 1990). By comparing calculated with actual levels of surface water, it is possible to conclude how much groundwater is being discharged or recharged. In other words, the combination of hydrological modelling and empirical data allows indirect estimation of the missing information on discharge and recharge.

In addition, hydrological modelling of groundwater flows can serve to predict alternative states of the groundwater system, namely, by formulating different scenarios, each of which is associated with particular boundary conditions. For instance, consider a wetland plain with a large recharge of groundwater from a nearby hill ridge (Fig. 2.1). A scenario approach could include boundary

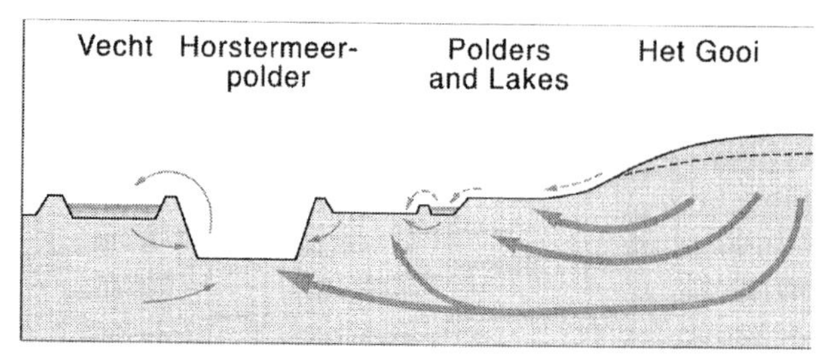

Fig. 2.1 A wetland plain with a large recharge of groundwater from a nearby hill ridge.

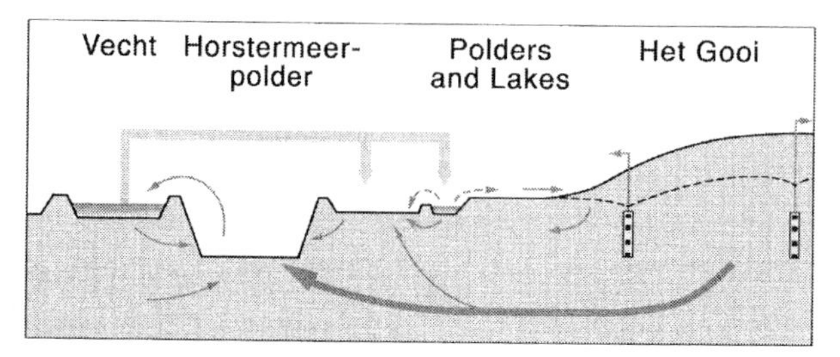

Fig. 2.2 A wetland plain with abstraction of drinking water from a nearby hill ridge.

conditions that reflect the abstraction of drinking water on the hill ridge. This would then allow assessment of the effects on groundwater levels, not only in the hill ridge but also in the adjacent wetland plain (Fig. 2.2). It is evident that a scenario analysis can also be used to assess the effects of water management, such as control of groundwater tables in polders or water abstraction levels.

Like groundwater quantity, groundwater quality or chemistry is also affected by the different water sources (i.e. by precipitation, groundwater from elevated areas, surface water from streams and rivers). All have specific, and generally distinct, chemical properties (Likens *et al.*, 1977; Schot and van der Wal, 1992).

- Rainwater is condensed water that contains few minerals. In many countries, notably the Netherlands, nowadays it tends to be characterised by high acidity.
- Groundwater is infiltrated rainwater, which often has been present in the subsoil for many decades or even centuries. As a result, it has reached a chemical equilibrium. Groundwater is well buffered, because of the high concentrations of calcium and hydrocarbonate in the soil, and is characterised by anoxic conditions: rich in iron and poor in nutrients. When groundwater seeps up, through discharge, the calcium and iron in it can chemically fix phosphates, resulting in nutrient-poor conditions.

- The water quality of many rivers and canals is affected by human activities in the stream area, such as nutrients and pollutants from agriculture, sewage systems and industry.

Since the sources of water in a hydrological system affect its chemistry, manipulation of these sources by policy and management is required to control or improve water quality in the system.

2.2.2 *Hydrological characteristics of wetlands*

Wetlands are an important component in the hydrological cycle because they can store large amounts of water for long periods of time. Wetlands are wet as a result of two possible factors. First, an impermeable layer in the soil, such as a heavy clay soil or an iron pan, can prevent water from leaving the wetland basin. Second, the inflow of water into a wetland can be larger than the outflow of water because of a constant inflow of groundwater, an input from river water or an extremely high level of precipitation. Freshwater wetlands are commonly found in depressions in the landscape, where water can flow in under gravity but cannot easily flow out again.

A large variety of wetland types can be found, depending on regions and climate types. Table 1.2 offered a classification of wetland types into 42 different classes. These can be grouped into six categories, according to the characteristic hydrology and the origin of the water:

- saline water with tidal influences along the coasts of oceans and seas;
- brackish water in deltaic plains, the contact zone of river and sea;
- freshwater in rivers and brooks, with seasonal fluctuations;
- fresh groundwater in plains, with rather stable conditions;
- rainwater in elevated parts of the landscape, constantly stagnating without any fluctuation; and
- hydrological conditions created by human endeavours, such as dams, irrigation or pumps.

Each of these groups will now be briefly discussed in order to illustrate their essential hydrological differences.

Saline wetlands

Tidal influences in wetlands at the borders of the oceans and seas result in flooding with saltwater. Only a certain number of species can survive the saline conditions of these intertidal coastal wetlands. Two specific wetland systems in this category are well known: salt marshes and tropical mangroves. Salt marshes consist predominantly of grass species that are tolerant to salinity.

They have a high level of productivity because of the regular addition of nutrients through tidal fluxes and the rich sediment layers (Adam, 1998; Cartaxana and Catarino, 1997; Gosselink and Maltby, 1990). Tropical mangrove systems form from tree species that have aerial roots to survive the extreme conditions present (Mitsch and Gosselink, 1993; Plaziat *et al.*, 2001). Both systems are important as breeding and nursery grounds for many saltwater species. Moreover, their huge production of biomass provides food for migrating animals such as birds. The hydrology of these systems is simple: they are flooded at high tide, while at low tide the water drains to the sea via tidal creeks. Usually, there is a characteristic gradient in salinity from the sea towards the land behind a marsh or mangrove.

Brackish wetlands

Deltaic plains of rivers are dynamic systems where river flows of freshwater meet and mix with tidally driven saline water from the sea. In most cases, there is a gradient from freshwater to saltwater (Mitsch and Gosselink, 1993). Estuaries display extreme dynamic patterns, not only in salinity but also in sedimentation and erosion. They are characterised by a predominance of brackish lagoons and intertidal mud flats, which tend to be highly productive. This highly productive wetland system provides breeding and nursery grounds for residents as well as transient, including migratory, species (Deeley and Paling, 1998; Weinstein *et al.*, 2001).

Freshwater riparian wetlands

The hydrology of the riparian wetlands adjacent to rivers is dominated by the flow of fresh surface water in the river. Seasonal fluctuations in water level make these systems very dynamic (Mitsch and Gosselink, 1993; Tockner *et al.*, 2000). At high river discharges, water levels will rise to inundate the river's floodplain, including the riparian wetlands. At low river discharges, both floodplain and wetlands may become very dry. The discharge is a function of the precipitation and temperature in the river basin. The fresh surface water tends to be rich in minerals and nutrients, which buffer the ecosystem with a medium acidity (Moss, 1988). Sediment transport by the surface water during high discharges causes coarse sediments of gravel and sand to be deposited close to the river and fine sediments of clay are deposited in the floodplain. As a consequence of variations in the seasonal patterns of flooding, sediment sorting and soil development, riparian wetlands show a large diversity in habitat. This is reflected in the distribution of different types of vegetation over the floodplain, such as reedbeds, grasslands, riparian forests and oxbow lakes (Ellery *et al.*, 2000; Verhoeven, 1992).

Fen wetlands

Fen wetlands are influenced primarily by groundwater, which tends to occupy depressions – the lowest parts – in the landscape. Rain falling on adjacent areas with a higher elevation infiltrates the soil and subsequently flows, very slowly, towards these depressions. The groundwater is rich in calcium, which, together with low concentrations of nutrients and a stable temperature, perfectly buffers the wetlands (Succow, 1988). The hydrology of fens is not very dynamic, causing them to remain wet even in dry periods. The characteristic vegetation is rich in species that are supported by the nutrient-poor water. Accumulation of organic material or peat may occur. Fen wetlands are sometimes also found in contact zones of river valleys or dune areas (Grootjans *et al.*, 1998; Wassen *et al.*, 1990).

Elevated wetlands

In elevated parts of the landscape, where sea, river and groundwater have no influence, the only way wet conditions can occur is through precipitation. The hydrology of the wetlands at these locations is characterised by a large surplus of precipitation over evaporation, combined with an impermeable soil structure so that the surplus cannot disappear as groundwater. Under these circumstances, bogs can be formed, which are characterised by waterlogged conditions. Like groundwater-supplied wetlands, bogs show stable hydrological conditions and low nutrient availability. However, the rainwater source promotes more acid conditions, which are not buffered. Moreover, bogs are poor in minerals. Vegetation typical of bogs, such as *Sphagnum* mosses, shows its optimum growth under these conditions. Because of the acidity, litter and the organic material from the vegetation cannot mineralise. This gives rise to the accumulation of organic material, resulting in peat layers. These can sometimes be several metres deep (Moore and Bellamy, 1974; Succow and Jeschke, 1986; Verhoeven, 1992). The hydrology of a bog can be regarded as that of a sponge, with the peat layer retaining most rainwater. Moreover, losses of water via evapo-transpiration are restricted by the climatic conditions. This causes the flow of water to be very limited. The larger the bog, the more effectively is rainwater captured and held in the system (Couwenberg and Joosten, 1999). Bogs are widely distributed in the boreal parts of the northern hemisphere (i.e. Siberia and Canada). The diversity in species in bogs is limited by the harsh conditions created by the combination of low pH, the absence of nutrients and the climate.

Wetlands affected or created by humans

Many wetlands in the world have been altered as a result of human-induced changes to local hydrology. Drainage and reclamation have converted a huge number of wetlands into agricultural or urban land. Rainwater and upward seepage of groundwater then force the use of pumps so as to maintain sufficiently low water levels. In northwest Europe, the hydrological conditions of large areas are maintained by pumps. Groundwater is close to the surface and surface water is abundantly present in the many canals and ditches, which affect drainage. According to the Ramsar definition of wetlands, such areas are still wetlands, even though the hydrology is no longer 'natural'. Remnant natural wetlands often remain as islands in an agricultural landscape. For instance, thousands of pot-holes are embedded in the agricultural landscape of northern America, where they support the diversity in waterfowl (van der Valk, 1989). Human activities have not only resulted in the removal of wetland systems, they have also created wetlands, such as ponds, lakes and irrigated land, for a variety of purposes. Some wetlands have developed incidentally with the completion of extractive processes, such as sand, coal and peat mining. Lakes have been created for such purposes as recreation, hydroelectricity generation, water storage and water purification (e.g. sewage treatment). In many cases, the water management is organised by dams or pumping engines. For example, in the Netherlands, water tables in wetland plains are artificially managed by pumping water surpluses out of polders (areas of low-lying reclaimed land). This water management is necessary to permit agricultural land use. In dry periods, polder water is sometimes supplemented with surface water from rivers or canals in order to facilitate the maintenance of constant water tables. Possible negative consequences of this water management are the mixing of waters with different chemical compositions, and the disturbance of natural patterns of water flow. Some of these elements will be discussed in more detail later in this book (notably Ch. 5), when we present a case study for the Netherlands.

2.2.3 *Soil characteristics and processes*

The conditions in the topsoil determine most of the chemical and biological processes in wetlands. Mineralisation processes, an essential part of the recycling of organic matter, are facilitated by the chemical conditions in the soil (Dominico and Schwartz, 1990; Scheffer and Schachtschabel, 1984). There are three sets of basic conditions.

Mineral composition In general, coarse sandy or rocky soils are composed of basic elements such as silica, and are poor in minerals. By contrast, clay soils, which have been subjected to weathering processes, contain various elements and are rich in minerals. These are then available for biotic processes. Sandy soils can store little water, whereas clay soils are effective stores.

Acidity Most biotic processes need neutral conditions. If the topsoil is acidified, for example from leaching processes or acid rainwater, many biological cycling processes stop. This results in a shortage of nutrients for flora and fauna. In many cases, clay soils contain a mixture of minerals that can buffer the influence of acidification.

Humidity All life needs water. If there is insufficient water, life cycles stop. Normally, humid soils contain sufficient water as well as oxygen to facilitate biotic processes. If soils become waterlogged (i.e. water tables are constantly so high that there is no oxygen in the soil) the cycling of dead organic material is retarded, which will lead to an accumulation of organic material (see the discussion of elevated wetlands in Section 2.2.2). If this process occurs over a long period of time, layers of organic material can accumulate (Succow, 1988). The result is known as peat. Under acid conditions and under the influence of rainwater, bogs are produced. Under more neutral conditions with buffered groundwater, fens form.

These basic conditions vary among landscapes, grading from a river or sea to more elevated – and, therefore, drier – areas (see Fig. 2.3). Surface water flows towards rivers or seas. It carries much natural waste and debris and so is rich in nutrients and minerals. Sedimentation of clay particles suspended in the water creates fertile, humid soil where it is deposited. The more elevated regions are dryer, as a result of leaching in the topsoil, and contain relatively small quantities of minerals and nutrients. Here, vegetation grows slowly and the area is relatively poor in species.

Between these two extremes (i.e. source and sink areas for both water and sediments), three different types of wetland can be found (Schot and Molenaar, 1992). First, close to the river, the influence of the buffered and nutrient-rich

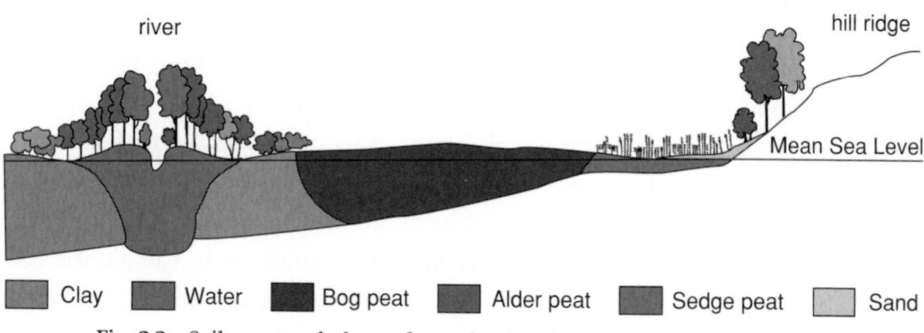

Fig. 2.3 Soil geomorphology of a wetland plain.

water, in combination with high water tables, induces peat accumulation and results in fens and marshes. Second, comparable conditions may arise in wetlands close to the elevated area, since the leaching process in the higher area results in a flow of buffered groundwater into bordering wetlands. Buffered wet conditions will prevail and, as the elevated area is relatively poor in nutrients, the wetlands will become poor in nutrients too. The result is a fen that creates unique conditions for characteristic species. Third, a zone lies between these two fen types, that is only influenced by acid rainwater. A bog can develop in the long term under such wet and acid conditions.

Far below ground level, the geomorphology also influences the wetland ecosystem. As discussed in Section 2.2.1, the flow of water varies with the resistance offered by different soil types. If the soil geomorphology includes different layers, there can be groundwater flows in the sand layers (aquifers), whereas the interrupting clay layers retard water flow (aquitards).

2.2.4 *Ecology*

Whereas the flow of water can be calculated relatively easily given straightforward physical processes, the interrelations in an ecosystem are far more complex as they are made up of feedback mechanisms, species interactions, biotic–abiotic interactions and spatio-temporal processes (Mitsch and Gosselink, 1993). An ecosystem includes all abiotic conditions and variables, alongside the biotic processes. The biotic component can respond to changes in abiotic conditions in an unpredictable way.

Each species is influenced by a set of factors, including water tables, chemistry, elevation and weather conditions. Each species requires a set of close-to-optimal conditions in order to live and grow. For instance, a number of species living in freshwater wetlands in nutrient-rich conditions cannot flourish in any other type of soil or under different chemical characteristics. Other species prefer peat soils and buffered conditions and show no special preference for nutrient concentration. Of course, there are also a number of specialised species that actually prefer nutrient-poor and buffered peat soils. These specialised species are rare because they are characteristic of specific conditions that seldom occur. The most common species to be met in ecosystems are usually those that require the less-specific conditions. Each species responds differently to the variables in the ecosystem; two extreme responses are ‘special species’, with a very narrow range of tolerance to variation, and ‘common species’, with a wide range of tolerance.

If a wetland is fairly stable and there are no fluctuations in the abiotic conditions, a large number of plant and animal species can live there. To prevent inbreeding and loss of genetic information, a large number of specimens of these species should be present. The number of specimens living together is called a population (Andrewartha, 1972). A population has one important biological task: to try to maintain the population with its genetic information. The number of seeds produced by plants, or the number of juveniles produced by animals, has to be high enough for the next generation to be present at least in equal numbers. The basic principle is that each species has a level of reproduction which enables that species to be present in the future with the same population size. In one year's time, the number of dead specimens ought to be replaced by new ones. This balance is influenced, however, by a large number of stochastic factors. For example, one extremely cold winter could cause half the population of a fish-eating bird such as a heron to die. This leads to a fall in the number of young herons since the number of pairs nesting the following spring will be half that in a normal year. Not only will the number of adult birds be lower one year later but eventually also the number of juveniles. The population size will fall at that point even if it can recover within a few years. The smaller number of birds will be able to find their food more easily use the best opportunities in the ecosystem; consequently, they will have more juveniles than in normal years. Fluctuations in populations are normal and are a characteristic of ecosystems subject to variations in abiotic conditions.

The fact that each species has its own tolerance to the environmental variables, and that the population in each species is fluctuating in numbers, is relevant to the competition between species. Each species needs to expand its population and is in competition with others. This is also true for the presence of light for plants. A dense and high vegetation hinders the plants living at surface level; consequently, they strive to grow tall to catch the solar energy. Two animal species can need the same prey species to survive and the one that is most competitive will find most food and will have the most young. An increase in certain nutrient levels will stimulate specialist plant species to grow faster, so that the space of others – not stimulated by those nutrients – is taken (e.g. van Duren *et al.*, 1997). This competition is always present, but since the ecosystem is fluctuating, the wheel of fortune constantly changes, favouring different animal and plant species at any one time (Huisman and Weissing, 1999).

Whereas flora plant species require solar energy and minerals, animals need plant and animal species as food. There are direct relationships between particular species, since feeding behaviour results in specialisation. A diagram can be drawn describing all the feeding relationships in an ecosystem. The question 'Who is eating whom?' can be answered with this network of relations, known as a food web. A simple example from a fen wetland (Fig. 2.4) illustrates how complex an ecosystem food web can be. In addition, many species are close to the base of the food web and only a few predators are at the top. This pyramid structure is

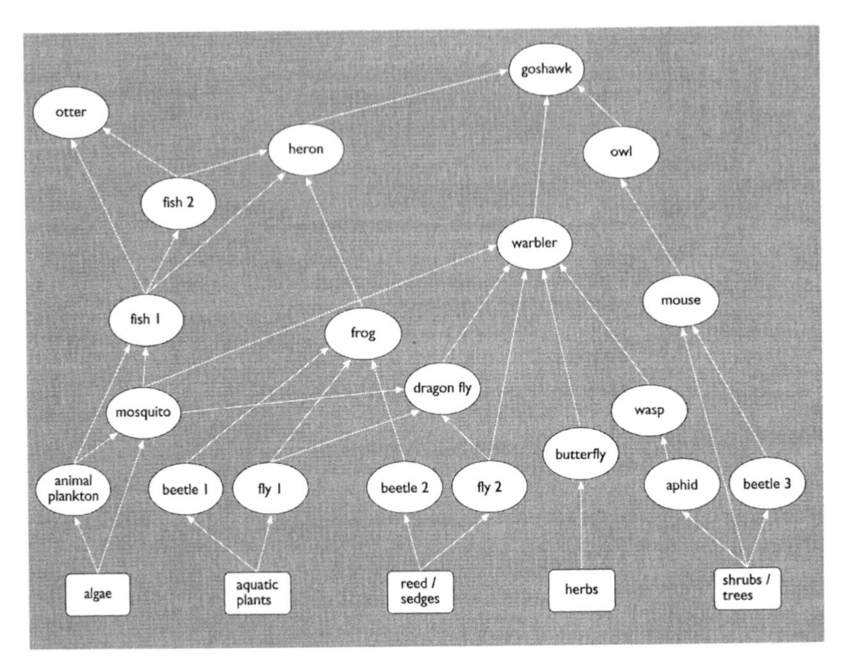

Fig. 2.4 The food web in a fen wetland.

also present in the quantity of biomass, where the top predators are represented by relatively small numbers.

To survive as a population, many species also have a characteristic life strategy known as specialisation (see Grime, 1979). If an individual plant species is to flourish, it has two possibilities for investing its energy. It can produce as many small seeds as possible and try to let these seeds germinate in as many locations as possible, or it can invest in a restricted number of seeds with extra spare food so that if they germinate they will survive for a longer period. Conditions fluctuate dramatically in dynamic pioneer ecosystems, and it is profitable to use each location for germination, since in the next year the conditions might be far worse. In this case, it is better for the plant species to produce as many seeds as possible and to spread these out over a wide area. By comparison, when the conditions are stable, it is to the plant's advantage to drop a smaller number of large seeds with extra energy. As it is the plant's strategy to inhabit stable conditions, the best chance of germination is given by providing the seed with a surplus of energy to ensure its growth.

These strategies are also present in animal life (Krebs, 1972; Odum, 1975). There are always 'optimistic' animal species that invest most energy in reproduction at the expense of their own lives, whereas other species invest primarily in their own survival, with reproduction seemingly having secondary importance.

When food is scarce, birds of prey (which have a long lifespan) do not build a nest in spring, as they know that the number of prey is too restricted and that their juveniles would get insufficient food. In this situation, they invest in their own survival to ensure that they will still be around to reproduce the following year. This is a more profitable strategy for the population as a whole.

A similar life strategy can be identified for some plant species. Some plants can live in extreme conditions with high acidity and very low nutrient levels, since they succeed in recycling all the nutrients necessary for the life cycle in the next year (Grime, 1979). In these conditions, they are far more competitive than other species that have a life strategy concentrated on using the nutrients directly but not recycling them in their roots. The latter are the 'big spenders'. As a result, they lose their nutrients and can suffer a shortage. However, in nutrient-rich conditions they will always win the fight for survival.

There are also many different types of specialisation in the animal kingdom to enable different species to survive the competition. For example, there is a difference in grazing activities between the main herbivorous mammals. Red deer and elk are browsers that eat the twigs of trees at about 1 m from ground level, whereas roe deer feeds on herbs and the lower parts of the shrubs and rabbits eat the lowest grass layer. These species can survive in the same area without competition. Other animals can survive in one area because they are separated by season, by the space that they inhabit or by their differing abilities to survive under extreme conditions.

Returning to the discussion about a wetland on an ecosystem level, it is necessary to consider the concept of succession. Succession is the trend to change the system from a pioneer phase to a climax phase (van der Maarel, 1980). The pioneer phase is characterised by extreme, unstable and dynamic conditions, within which only some species are adapted to dominate the ecosystem. During the succession, the fluctuations in the system are largely reduced and the relationships become far more complex.

A mudplain shows the characteristics of the pioneer phase. It can be flooded (e.g. five times a year) and only very 'optimistic species' will survive these conditions. Succession will change the chemical and biological conditions. After a phase with herbs, tree species can germinate and they will stabilise all kinds of process. As they cover the soil, extreme climate conditions are reduced. The production of dead leaves at the end of the season will result in an organic layer on the soil. Within this organic layer, conditions are buffered. As a result, many insect species can hide, humidity is more stable and the recycling of minerals and nutrients is ensured via bacteria, fungi, insects, etc. Moreover, more species inhabit the climax phases than the pioneer phases, so that a fluctuation within a pioneer phase will influence a relatively restricted number of species.

Finally, these ecological relationships show a distinct spatial pattern. This is influenced by the heterogeneity in the landscape, resulting in different abiotic conditions, and by the size of the populations. The larger the ecosystem, the bigger the population will be and, as a result, the less the population can be influenced by stochastic processes. For this reason, large wetlands are always to be preferred to fragmented ones. Populations split up into spatially separated subpopulations can still be in free contact by way of ecological corridors and are for that reason still viable, since they interact as one total meta-population consisting of a number of interacting subpopulations (Hanski, 1998). This is valid with one proviso: there need to be possibilities for the specimens to migrate within the landscape. The quality and the length of the ecological corridor determine the possible migration of the species to the other subpopulations. For this reason, the whole matrix of ecosystem elements and their active relationships are determined by the actual connection of the population within the landscape. Clearly, isolated patches of wetlands are less able to facilitate species diversity than interrelated ones.

This results in a very complex relationship between the populations of many species and the conditions in the wetlands. Each phase in the succession has its own set of species; however, during the succession the number of species will increase. If stress factors, such as flooding, acidification, eutrophication or salinity, increase, other species will be stimulated in the competition for space, light and settlement (Bornette *et al.*, 1998). Alongside these changes, there are feedback mechanisms. If the nutrient levels increase, specific species that fix the nutrients in their tissues will be stimulated and the nutrient levels will be reduced. If there is an increase in turbidity in surface water, a dense aquatic vegetation will create calm conditions between the leaves and, as a consequence, the turbidity will settle down. If one fish species is dominating, one year later the number of predators will increase too and will reduce the number of the dominating specimen of this fish.

Such stabilising or homeostatic processes can be found in many ecosystems. The ecosystem is maintained for as long as possible by chained feedback mechanisms, until the final threshold is reached and the system is disturbed (Scheffer *et al.*, 1993). For instance, if the aquatic vegetation dies, resulting in a sudden increase in turbidity and algal bloom, even the nutrient-rich type of vegetation will not be able to receive solar energy from light. In this situation, fish species that like to grub around in the sediments are stimulated. By their activities they maintain the turbidity and the carnivorous fish species find it harder to see their prey and will starve. In this manner, the number of species of fish that grub is not reduced any further and the population will grow, with ensuing negative effects on the wetlands. If there is a high level of water pollution present in the system, the problems will increase. This sequence is also valid with other variables for terrestrial systems. For example, a high level of biodiversity can change into a monoculture if a plant species that induces increased litter fall becomes dominant, covering the soil so that no other species can survive. For a very concise introduction to ecology for environmental scientists and economists, see Folke (1999).

2.2.5 *Functions of, and processes in, wetlands*

Soil, water, chemical and biological features are all relevant to the description of the functions and processes in wetlands. We will discuss some examples to illustrate the integration of these four aspects in wetlands.

The physical functions are well known, for instance in coastal protection. Erosion by waves can cause shorelines to retreat, resulting in loss of property or in the flooding of a large area. Coastal marshes absorb wave energy and in this way protect the coast from erosion. Within 30 m, most of the wave energy is eliminated by the marsh vegetation (Schot, 1999). Mangrove swamps in the tropical regions provide excellent protection during storms. At the same time, these swamps perform another function, by creating new locations for settlement. The mangrove trees, with their large root system, trap sediments, resulting in extra protection against flooding. When fast-flowing water carrying sediments reaches a shallow wetland area, the velocity is reduced and the sediments can be deposited. The effect of the laws of physics upon soil are relevant to this situation.

An important hydrological function of wetlands is that they maintain an equilibrium between the recharging and the discharging of aquifers. Groundwater and surface water are strongly interrelated. The groundwater table under the land surface has a continuous interaction with the surface water tables in nearby lakes. Depending on the surface water tables, the wetlands perform discharge or recharge functions for groundwater aquifers. When surface water tables are low, a discharge of groundwater can supply the wetland; since this flow is mostly stable and slow, continuous wet conditions are provided. When surface water tables are very high (for example after flooding events), wetlands may recharge groundwater through infiltration. People living at great distance from the wetland may benefit from this process as they can remove groundwater to provide drinking water and irrigation water for agricultural purposes.

Because of the need for transportation, most human occupation first established close to rivers. Exceptional weather conditions, with high rainfall, storms or fast-melting snow, can cause high discharges of water. The wetlands can store run-off water that originates in other areas, without damage to the ecosystem. The species that inhabit these wetlands are selected by their ability to survive these conditions. In this way, the wetlands reduce the peak flow, prevent flooding and protect human occupation. Therefore, wetlands perform a hydrological function by providing a required water-storing capacity for rivers.

Wetlands can play a major role in chemical processing. The eutrophication and pollution of surface water are problems present in many countries. Wetlands are able to reduce the amount of nutrients and pollution in an area. This is one of the chemical functions of wetlands. Together with sediment trapping and flocculation, precipitation and absorption are important processes in storing great quantities of chemicals from surface water. Many nutrients, heavy metals and pesticides are removed from the water and settle in anaerobic sediments or are used in biological processes through uptake into plants and algae. This sanitation function of wetlands is often produced artificially by establishing purification beds in new wetlands.

The chemical function of wetlands has an important role in the cycling of carbon, oxygen, nitrogen and sulphur (Mitsch and Gosselink, 1993). The waterlogged conditions favour bacteria that reduce the oxidised forms of carbon, nitrogen and sulphur. Together with decomposition by algae and fungi, such bacteria enable wetlands to be among the important ecosystems on earth for recycling essential chemical components. In addition, wetlands can be an important carbon sink through the fixation of carbon dioxide by wetland vegetation, thus countering the enhanced greenhouse effect. As decomposition processes are delayed by waterlogging, undisturbed wetlands can accumulate organic plant remains by peat formation.

The abiotic conditions that prevail in a wetland are important for its role in chemical and biological processes (Wassen, 1995). Domination of acid rainwater results in increased growth of bog vegetation. Domination of calcareous groundwater results in the fixation of phosphorus and other nutrients; this creates nutrient-poor conditions and biomass production is restricted. Nutrient-rich surface water will stimulate the rapid growth of algae, so less oxygen will be available to other species. The activity of the bacteria, algae and higher plants are influenced directly by chemical conditions such as acidity and salinity. Therefore, characteristic chemical processes in wetlands, such as carbon fixation, sulphur recycling and nutrient fixation, are inextricably linked with hydrological conditions, chemical equilibria and biological growth.

A major aspect of the biological function of wetlands is their high productivity. Only some tropical forests and the most intensively cultivated areas in the world have a higher production. Approximately a quarter of the world's net primary production occurs in wetlands, which form only 6% of the land surface (Costanza *et al.*, 1997a,b). This high production can occur because of the constant availability of water and nutrients; the latter imported by water from other areas.

With this high productivity, wetlands are very efficient converters of solar energy to fixed carbon and biomass. The high productivity and the many niches available in this aquatic and terrestrial system gives rise to the presence of many invertebrate species, such as insects. This enables many birds and mammals to find large quantities of food. Some animals use the wetlands in specific phases of their life. Spawning fish use the shallow water of the wetland and migratory birds look for the locations with the highest potential for food during their migration. Large areas of wetlands, which are typically relatively undisturbed areas, provide many facilities for the settlement of a variety of species.

However, this set of profitable wetland functions can be disturbed in many ways, even if the wetland ecosystem is very vigorous and can itself solve many problems caused by human activities. Stable conditions, with high production in biomass, carbon fixation, flood attenuation and high biodiversity, can deteriorate. The ecosystem will collapse if, for example, the vegetation is flooded for a long period of time as a result of a change in input of groundwater or rainwater. Many life cycles would then no longer be possible. Too much water will also result in high water-flow velocity, causing erosion and a lower rate of sedimentation. Moreover, nutrients will be released and transported to other areas. The overall result will be a disturbance of all aspects of the wetland.

Disturbance can also result from the opposite scenario: too little water available for the ecosystem will remove the characteristic wet conditions. This will adversely effect the characteristic processes and species, which cannot be maintained in dry conditions. Reclaiming a wetland, or using all the groundwater for other reasons, will directly influence these processes. If organic soils are drained too much, mineralisation of the peat layer will result and carbon and nutrients will become available. This will give rise to a loss of the carbon sink function and the eutrophication of the wetland. Other causes of disturbance are extreme eutrophication and pollution with heavy metals or pesticides.

In the case of mangrove swamps, sedimentation and the cycling of elements are connected to a high level of biomass, as well as to a variety of processes (Gilbert and Janssen, 1998). A broader ecological function is that the mangrove swamps enable many animals to use the surplus nutrients and food, and that they provide cover, allowing predators to hide. Mangrove swamps are the essential spawning grounds of many species, which often live far away from these wetlands.

2.2.6 *Ecological evaluation of wetlands*

Nature can be given many values, both instrumental and intrinsic. In order to assess the relative value of natural areas, comparative ecological evaluation – sometimes referred to as ecological valuation – studies have been performed (Edward-Jones *et al.*, 2000; Nunes *et al.*, 2002). These have focused on five main categories of issue: species, diversity, ecosystem, human threats and human benefits.

Species

Valuation on a species level has been based on population ecology (Andrewartha, 1972). An important question here is which minimum population size assures viability and stability of the respective population. The answer evidently varies from species to species. A certain population size can be sufficient for the viability of a large mammal, whereas the same number may be too small to sustain a viable population of mice. The first aspect to consider is whether or not there is a population that can survive into the future. The second aspect to consider is the importance of location in the distribution of the species. Common species are regarded as less important, while the rarer the species is, the greater importance is given to its presence. On a national or even an international level (IUCN, 1993), this scarcity is described in 'red lists', which contain the critically endangered, endangered and vulnerable species from a defined group. Three elements are important in defining these red lists: the reduction of the population during recent times, the scarcity of the population in a geographic area (e.g. country) and the international importance of the presence of the population. Another aspect in the valuation can be the importance of the presence of the species to the function of the ecosystem. An extreme example of the importance of population size is where a species is endemic to only one location (usually an island or a mountain range) that constitutes the worldwide distribution of that species (Simberloff, 1998). Sometimes ecological evaluation at this level fails through lack of information regarding the number of stable populations of a particular species and the distribution of the species in the region, the country or the world.

Diversity

Diversity is often considered in the valuation of natural areas (Gaston, 1996, 1998). In many cases, a rich diversity of species is an indication that the conditions are optimal and preservation is in order (Spellerberg, 1996). If we protect all areas that have a high number of species, most of the species native to the whole country will be preserved. This is the starting point in the UK conservation policy with the Sites of Special Scientific Interest (English Nature, 1993). A quick and simple evaluative method is to count the number of species that are present in one area and to compare this with the number inhabiting other similar areas. This requires an examination of all the plants and animals that are present, since missing values influence the assessment. The total number of species might be substituted by the number of characteristic species, or by the number of dominant species. Incorporation of these numbers into the populations as a whole can indicate the natural structure of a certain area. A long list of species that are seldom found might indicate a rare condition but could also indicate an area where most populations have almost disappeared. Alongside the simple calculations, a number of aggregate indexes are available to evaluate the numbers, such as the Shannon index, the Simpson index and the Evenness index (Hutcheson, 1970; Simpson, 1949; Odum, 1975, respectively).

Ecosystem

The ecosystem can, just like the species, be assessed in terms of rarity of being observed in a country or continent (Bibby, 1998). There are, however, other aspects. The area covered by a wetland is important, since generally smaller wetlands will contain fewer species. The absence of ecological connections between wetland areas is a measure of their isolation. Linked to this, most natural processes, resulting from undisturbed characteristics, that are highly valued can only emerge when areas are large enough. In essence, this is an issue of the degree of naturalness of a wetland. By defining how the wetland should be, a reference situation with characteristic processes and species is created. We can use this reference to identify the number of characteristic species, the number of accompanying species and the absence of other species, or various ecosystems.

The last two items consider the human impact upon wetlands.

Human threats

There is the need to evaluate the threats that humans pose to wetlands. Some wetlands are quite robust, that is the environmental pressure caused by human activities has little or no influence on their processes and functions. The conditions in many other wetlands, however, can be easily disturbed by relatively simple changes. Such sensitive wetlands are often highly valued, as is indicated by their preservation being costly in net terms, taking opportunity costs – of alternative land use – into account. A concern for preservation and conservation can be followed by an evaluation of options for restoring a disturbed wetland. One can examine, for instance, how easy it would be to facilitate a new settlement of species, and whether or not the potential of the site would justify the effort required.

Human benefits

The last level of ecological evaluation is a purely human level, not directly connected with the natural conditions in the wetland (Edward-Jones *et al.*, 2000). Three aspects are relevant. First, the historical setting in the wetland area can be regarded as important. For instance, a drainage system might have been present for several centuries, thus providing a good picture of the cultural heritage of the combined artefact-ecosystem. Second, the extent to which wetlands contribute to education can be examined. For example, each time a wetland is visited by people interested in natural processes, society's awareness of the importance of wetland areas will increase. Finally, a wetland can be important for scientific purposes, which applies to both the natural and the social sciences. In particular, a variety of wetland areas, in terms of processes, functions, human activities and public regulation, can serve as a basis for understanding possible links between natural and social–economic systems.

2.2.7 *Assessment of environmental quality*

The concept of ecological or environmental quality attempts to define the potential of ecosystems (in the case study of this book, wetland and mixed agricultural–wetland ecosystems) to support human welfare (see Ch. 5 onwards). Environmental quality, as an objective for the evaluation of alternative scenarios, is defined as an ecosystem that is functioning well.

Various terms have been used to describe how well a natural system such as an ecosystem is functioning or, conversely, how disturbance has affected its functioning. The main ones are ecological sustainability, biological diversity, ecosystem health and ecological integrity. Their definitions and characteristics are listed in Table 2.1.

The literature mentioned in the table suggests that there are three characteristics of ecosystems that relate to their ecological functioning:

- processes within an ecosystem;
- ecosystem structure; and
- ecosystem resilience.

Ecosystem process refers to ecosystem activity as reflected by productivity and the (re)cycling of carbon and nutrients. *Ecosystem structure* relates to species composition and community characteristics, as reflected, for example, by biological diversity. *Ecosystem resilience* refers to the ability of ecosystems to maintain structure and pattern of behaviour in the presence of stress (Holling, 1986). It consists of two elements: the time taken to recover after disturbance (Pimm, 1984) and the amount of disturbance required to move a system out of one state and into another (Holling, 1973). The perspectives of each of the concepts from Table 2.1 on these three ecosystem characteristics are summarised in Table 2.2. (For further discussion, see Section 9.2.1.)

Further specification of ecological quality and of its components may be found in the context of the case study in later chapters of this book.

Table 2.1 *Definitions and characteristics of concepts used for the assessment of (disturbed) ecosystems*

Concept	Definition	Characteristics
Ecological sustainability (IUCN, UNEP, WWF, 1980; van Ierland and de Man, 1993)	Guarantee of long-term functioning, conservation of biological diversity and quality and quantity of natural resources	Carrying capacity not exceeded; no loss of species or ecosystems; extraction rate < regeneration rate; pollution rate < assimilation capacity
Biological diversity (Ghilarov, 1996; Harper and Hawksworth, 1995)	The variability among living organisms, including diversity within species, between species and between ecosystems	Distinction between genetic, organismal and ecological diversity, at the global level: species extinction; risk for life-support functions; adaptation through intra- and interspecies selection Distinction at the ecosystem level: diversity as a measure for ecosystem condition; relationship with ecosystem processes.
Ecosystem health (Constanza *et al.*, 1992; Mageau *et al.*, 1998; Rapport *et al.*, 1998, 1999; Schaeffer *et al.*, 1988; Steedman, 1994)	Hierarchical multiscale, dynamic measure of system vigour, organisation and resilience	Human health perspective for treatment and diagnosis by abnormal signs; propagates combination of scientific methods and societal references Healthy system has a balance between: vigour (productivity, activity, metabolism), organisation (number and diversity of interactions of components) and resilience (ability to maintain structure in the presence of stress), consisting of maximum magnitude of stress and recovery time
Ecological integrity (Karr 1981, 1991)	Ability to maintain a balanced, integrated, adaptive community of organisms having a species composition, diversity and functional organisation comparable to that of a natural habitat of the region	Structure: physical organisation of elements Composition: identity and variety of elements Function: operation of ecosystem processes Biotic integrity: index based on unimpacted regional reference site

Table 2.2 *Perspectives of different concepts of disturbed ecosystems with regard to the three ecosystem characteristics of ecological functioning*

Concept	Ecosystem process	Ecosystem structure	Ecosystem resilience
Ecological sustainability	Long-term productivity	Conservation of biological diversity	Carrying capacity
Ecosystem health	Vigour	Organisation	Resilience
Biological diversity	Functional guilds, keystone species	Genetic, organismal and ecological diversity	Adaptation by species selection; resilience by overlap of functional groups
Ecological integrity	Function	Structure and composition	Ability to maintain adaptive community

2.3 Environmental and ecological economics

2.3.1 *History and scope*

Environmental economics has developed as a branch of economics that is concerned with the economic analysis of the causes and the nature of environmental problems and their solutions. This includes issues relating to markets as well as to public policy. Environmental economics is often interpreted to include resource economics. This makes sense, since resource issues are intricately linked to environmental issues, an insight that has made the notion of sustainable development become a pillar of modern environmental economics.

Environmental economics developed during the 1960s out of applied welfare economics and was influenced by insights from, and approaches to, agricultural and resource economics. Nevertheless, long before that time, economists had shown an interest in environmental issues, such as those related to agriculture, population and food production (Malthus), productivity of land (Ricardo), depletion of coal stocks (Jevons), broader categories of depletable resources (Gray, Hotelling) and externalities and environmental taxation (Pigou). The early development of environmental economics during the 1960s and 1970s was dominated by three themes. First, cost–benefit analysis was applied to investment projects with environmental impacts, and, in line with this, monetary valuation techniques were developed and applied to assess the value of (or lack of) policies and projects that cause environmental changes and damage. Second, environmental policy theory was developed, aimed at a comparison, design and evaluation of environmental policy instruments. Third, mainly in response to the first report to the Club of Rome (*The Limits to Growth*; Meadows *et al.*, 1972), economic growth and resource scarcity were examined in theoretical and empirical studies.

Environmental economics uses concepts and models from neoclassical welfare theory (microeconomics). Its core insights are, therefore, critically dependent on the assumptions of rational individual behaviour (utility or profit maximisation) and market clearing, together guaranteeing an economic equilibrium: that is, unique combinations of prices and traded quantities of products on markets. Recently, during the 1990s, 'ecological economics' has developed as an alternative, broader approach. It functions as a forum for multidisciplinary environmental research in which economics plays an important role. The main differences between environmental and ecological economics are discussed in Section 2.3.3.

Closely related to environmental economics is resource economics. This covers a number of issues including indicators of resource scarcity, optimal resource extraction, imperfect resource markets, extraction of non-renewable resources (fossil fuels, metal ores, minerals) and use and management of renewable resources (water, forestry, fisheries).

Finally, an area known as 'environmental (business) management' has strong links with environmental economics, since it adopts a business or firm organisation perspective in order to understand firms' responses to environmental problems and policies. This field of research covers typical business administration topics, such as environmental care systems, environmental strategies, internal organisation, environmental accountancy, environmental reporting, environmental cost accounting and green marketing. So far, the textbooks have not integrated environmental economics and environmental management, which is consistent with the fact that, so far, these fields have developed rather independently.

2.3.2 *An overview of themes addressed by environmental economics*

This section provides a very brief summary of core themes in environmental economics. These include environmental policy theory, international environmental problems, growth and sustainable development (environmental macroeconomics) and monetary valuation.

The economic theory of environmental policy starts from the concept of 'externality' (or 'external effect', or 'external cost'), which can be defined as the influence of one economic agent's decision on the utility (welfare) or production of another agent that occurs outside the market and remains uncompensated (Baumol and Oates, 1988). The presence of externalities means that individuals do not have complete control over the set of factors that determines their production or utility levels. Environmental economics is particularly interested in negative environmental externalities: that is, negative physical effects of environmental pollution, resource use or other types of environmental disturbance of one agent by another. Ecosystem changes are considered relevant within the externality framework as long as they are caused by one economic agent while they have an impact on the welfare or production activities of other economic agents. An example in the context of wetlands is water abstraction for drinking water purposes, which lowers water levels in a wetland, thus reducing its species diversity and attraction for recreation. Externalities have been analytically examined and elaborated with the help of partial and general equilibrium theories. These are consistent with the earlier-mentioned neoclassical assumptions regarding individual behaviour and operation of markets.

A recent alternative, and both more dynamic and more ecological, perspective on environmental economics is provided by the notions of environmental sustainability and sustainable development. These focus attention on the stability of ecological systems' support of economic and human activities, dynamic processes and long-term growth, and equity and future generations. The dominant but also much criticised economic theory to elaborate these notions has been growth theory (Beltratti, 1996; van den Bergh and Hofkes, 1998).

Instruments of environmental policy are traditionally evaluated in economics on the basis of their efficiency features (Baumol and Oates, 1988). Effectiveness and distribution (equity) affect function as secondary evaluation criteria. The most common (archetypal) comparison is between uniform standards and taxes in pollution control. Taxes are considered attractive as they provide better incentives than standards when attempting to change individuals' behaviour and thus realise more efficient outcomes: either social welfare is higher or the costs of realising fixed targets are lower. The best known price instrument of environmental policy theory is the optimal or Pigovian tax, defined as equal to the marginal external costs (for example of pollution) in the optimal (hypothetical) equilibrium. This optimal tax is adapted for imperfect markets, dynamic contexts (technological innovation) and transaction costs. Standards are especially attractive from the perspective of effectiveness (or uncertainty). A combined instrument is a system of tradable permits (Tietenberg, 1996). This has two features: a ceiling is set on all pollutive emissions of a particular type by granting a finite amount of emission permits and such permits are tradable. The first feature ensures that the total (national, regional) emission level is restricted; the second feature assures flexibility and efficiency at the level of individual agents, with the ideal result that the costs of pollution abatement at the margin are equal among all firms (and equal to the equilibrium permit price). So far, this instrument has mainly been applied to controlling air pollution in the USA.

Recently, environmental tax reform has been studied in a wider context, particularly in examining whether it is true that a shift of the general tax basis from labour to environment (resources, energy, materials) would create a 'double dividend': less pollution and more employment. Theoretical analyses indicate that there is little reason for optimism here as environmental taxes will lead to a shifting of costs to the least-mobile factor, which usually is labour. It seems that the double dividend only has a chance when income distribution is allowed to become more uneven. So far, equity has received relatively little attention in environmental economics, although it is clear that equity considerations can act as a serious barrier in political decision making regarding stringent environmental policies (de Mooij, 1999; O'Riordan, 1997).

Since the late 1980s, in the wake of the popularity of the notion of sustainable development, international dimensions of problems and policy have received much attention in environmental economics. A first issue here is how to design environmental policies in open economies. This depends on the scale of the problem: local (solid waste, urban pollution), transboundary ('acid rain', river pollution) or global (greenhouse gas emissions and climate change). Another consideration is whether pollutive goods are traded. Many theories have been employed to clarify the link between foreign trade, environment and policy: partial equilibrium models, imperfect competition (firms with market power), strategic trade policy by countries, general equilibrium models, and statistical–econometric analyses (see van den Bergh, 1999, Part IV). Special attention has been given to the impact of policies, or lack of them, on developing countries. Some authors have claimed that the WTO (World Trade Organization) should be integrated with international environmental agreements, and that traditional comparative advantage theory no longer holds, given that all production factors (capital and labour) are immobile (Esty, 1994; Patterson, 1992). Transboundary environmental problems have also received much attention. Since the early 1990s, these have included global environmental issues, notably global warming, stratospheric ozone and biodiversity. Climate change has subsequently attracted a great deal of interest from economists, both in policy analysis and in modelling (see also Ch. 3). International agreements have been examined using game theory models of environmental conflict, agreement formation and agreement stability.

Related to international issues are spatial ones (see van den Bergh, 1999, Part V), which are of particular relevance to wetland studies. Aside from welfare economics, other sources of motivation and inspiration for environmental economics have been agricultural, regional, transport and urban economics. The relationship between land use and environmental quality seems essential for a good understanding of many environmental problems related to water, soil erosion, fragmentation of habitats, etc. Nevertheless, traditional economics and environmental economics are dominated by models and theories that assume spatial dimensions away (i.e. do not take space explicitly into account). In some recent research, space and location are regarded as important dimensions in the analysis of environmental quality and policy. For example, non-point source pollution control issues have been studied at the boundary of agricultural and environmental economics. Information asymmetries between polluters (farmers) and regulator are common here. Other spatially oriented research has close links with research on international trade and environment. Finally, transport has been much studied, as it is clear that it has significant impacts on both the environment and on congestion. Transport is characterised by the dispersion and mobility of sources, which require specific instruments. Theory suggests convincingly that a combination of road pricing – based on information about the location, car type and road congestion – petrol taxes and technical measures (motor size, speed) seems the most attractive policy approach. So far, this policy has received little political support (Button and Verhoef, 1998).

Environmental macroeconomics is a term referring to the macroscale of environmental issues, involving especially the impact of economic growth and the goal of sustainable development (see van den Bergh, 1999, Part VI). The environment has received much attention in macroeconomic modelling because economists have long been asked to provide information about the national economic consequences of environmental policy plans (see also Ch. 3). Growth theory has often been used to clarify relationships between growth, resources and pollution. At the moment, research in this area is dominated by endogenous growth theory, with relatively few new results so far. The growth debate has been active since the 1960s, but there has not as yet been any convergence among the various stances. The different views can be traced to differences in ethical stance regarding nature and future generations, and differences in assessment of the enormous uncertainties: for instance, regarding the long-term effects of decisions taken now and the stability of ecosystems in the face of human pressures.

An assessment of the growth debate suggests that a number of stances can be identified (van den Bergh and de Mooij, 1999): the *moralist*, who does not believe in endless progress; the *pessimist*, who does not trust the stability of ecosystems and availability of resources; the *opportunist*, who regards sustainable development as impossible anyway, and wants to enjoy the earth's natural resources for as long as possible; the *technologist*, who thinks that technology will make it possible to escape the environmental crisis; and the *optimist*, who even regards economic growth as essential for sustainability, as it changes preferences in favour of the environment ('environment as a luxury good'). The latter two perspectives have been examined empirically, which has given rise to the well-known environmental Kuznet's curve hypothesis: a bell-shaped relationship between environmental pressure and income per capita. The testing of this hypothesis has, however, not delivered substantial evidence that growth solves environmental problems. Instead, it seems that only local, health-related problems are solved or 'exported', and that problems distant in space or time (global warming) are neglected. One can explain this with the notions of spatial and temporal discounting.

Economic valuation is an area of research where environmental economists have really generated new economic theory (Freeman, 1993; Hanley and Spash, 1993). The basic idea here is that monetary values can be derived from individual behaviour or choices. Wetlands have been subject to many valuation studies (Brouwer *et al.*, 1999; Gren *et al.*, 1994; Nunes and van den Bergh 2001). The idea of valuation theory is that monetary valuation can be performed by comparing two states. In other words, monetary valuation applies only to changes, preferably relatively small ones. This means that the valuation of ecosystems as such is difficult, and instead valuation should start from a specific ecosystem-change scenario. A much noted study that tried to assess the monetary value of all ecosystems of the world, published in *Nature* (Costanza *et al.*, 1997a), is inconsistent with the basic assumption of monetary valuation theory. This is supported by various critical responses (see Pearce, 1998; and various invited contributions in *Ecological Economics* 1998 (25(1)). Apart from many other criticisms, the main problem is that this study lacks a realistic change scenario. The loss of all ecosystems of the world is not something an individual can value, if only because the individual will no longer be around under such an extreme situation.

Monetary valuation is performed using several concrete methods. Travel cost models try to assess the value of nature areas or parks by linking recreation demand to the generalised travel costs, including loss of time. Hedonic price models provide a relation between property (land, houses and sometimes wages), on the one hand, and environmental indicators on the other. Both these methods are said to 'reveal preferences': that is they use information about actual choices made by individuals in existing markets to assess non-market values for environmental goods and services. Another method, contingent valuation, has received the widest attention since it can be applied relatively easily. It is based on stated preferences: asking individuals directly for the value they would assign to a hypothetical change scenario. The main advantages of this method as opposed to the other two methods are that it can address a wide range of (hypothetical) environmental changes, including those that have not yet occurred, and that it can assess non-use values. The main disadvantage is that, because of its hypothetical nature, it suffers from various biases. In addition, the method is very sensitive to design characteristics: type of interview (face-to-face, mail, phone), scenario presentation (text, photo, film), amount of information presented and value elicitation procedure (starting values, guiding people, direct question). Economists have debated the correct procedures for this method, especially following valuation studies done in response to the Exxon-Valdez oil spill in Alaska in 1989 (see Arrow *et al.*, 1993; Carson *et al.*, 1992; Hausman, 1993). Finally, other methods for valuing environmental change can be based on production function techniques (notably in agriculture), and adopting theoretically less-attractive measures like defensive expenditures, relocation costs, restoration costs and opportunity costs.

Since the 1990s, interest has increased in two related methods: namely, meta-analysis (van den Bergh *et al.*, 1997) and value (benefit) transfer (Brouwer, 2000). Meta-analysis is the formal synthesis of results and findings of scientific studies. It can be used to summarise relationships and indicators over a collection of similar studies, to compare different methods applied to similar questions and to trace factors responsible for differing results across similar studies. Meta-analysis, and its focus on rigorous synthesis, contrasts with the more conventional literary review procedures. It can help to remove some of the subjectivity from analysis and forecasting, or at least make judgements more transparent. Value transfer, also known as benefits transfer, is aimed at transposing monetary values from one site to another. This is regarded as cost effective relative to performing a new 'primary valuation study'. Different options are available to do a value transfer: use an estimate from the best or nearest study; use the mean of estimates from various near studies; derive a regression model from a near study and predict (i.e. adjust parameters); and perform a meta-analysis on the basis of many similar studies and predict on the basis of the resulting 'value function'. The last is the ideal approach and this shows the important linkage between meta-analysis and benefits transfer. Transfer is controversial and to be effective it should satisfy a number of criteria: adequate data, sound economic methods, estimation of a willingness-to-pay function, similarity of populations of origin and destination sites and similar site characteristics (notably size). The world ecosystems' valuation study (Costanza *et al.*, 1997a), mentioned earlier, was in fact a value transfer study, but it did not satisfy these conditions. Value transfer can be used in principle to take wetland values assessed in many studies, undertaken mainly in the USA, the UK and Sweden, and transfer these to other regions, especially in a European context, with similar levels of development.

Economic valuation can be criticised from many angles (see Blamey and Common, 1999; Gowdy, 1997; van den Bergh *et al.*, 2000). First and foremost, the standard assumptions of economic valuation methods can be criticised, on the basis of notions of non-utilitarian altruism, ethical attitudes and 'citizen responses'. The last relates to the view that an individual sometimes acts as a consumer who maximises his/her utility and sometimes as a citizen who makes social choices, including those relating to environmental and nature (the consumer/citizen dichotomy). In addition, ordinary utility functions may be non-existent, because of lexicographic preferences, responsibility considerations and moral satisfaction. Lexicographic (hierarchical) preferences mean that substitution between arguments in utility is impossible. One can think here of the well-known Maslow pyramid (in order of priority: drink, food, clothing, housing, social contacts, self-esteem, love). Lexicographic preferences exist in relation to environmental goods and services that impact on human welfare when the services or goods they generate cannot be substituted by market goods and services. In this case, standard valuation theory is inapplicable. Finally, valuation depends on income so the estimated values will include a bias created by (national and international) income and welfare distribution. This does not, however, pose such a fundamental problem if values are simply used in a limited geographical area or in any other mainly economically homogeneous context. Section 3.4 will discuss the relationship between integrated modelling and monetary valuation.

Aside from valuation techniques, various other methods and models can be found in environmental economics (see van den Bergh, 1999, Part IX). Input–output analysis has been applied to sector structure in relation to polluting emissions. Computable general equilibrium models have been used to study the economy-wide impacts of the larger changes in environmental policy, which are expected to involve various significant price effects and substitution processes. On a theoretical level, game theory has become very popular in examining international policy choice and coordination. Optimal control theory has been traditionally used to study dynamic optimisation in resource allocation (both renewable and non-renewable), as well as economic growth (capital accumulation) with resource or pollution. Recently, decomposition methods have been employed in energy demand and environmental analysis to trace factors of change, such as volume (demand), sector structure, substitution, technical change, etc. This has also been combined with input–output tables, leading to structural decomposition analysis. Finally, advanced statistical methods have been applied in valuation research and, more recently, experimental techniques in 'economic laboratory' situations have been employed.

2.3.3 *Ecological versus traditional environmental economics*

In addition to the considerations involved in environmental economics, discussed above, ecological economics, a more recent approach, has paid close attention to ecosystem studies, including those relating to wetlands (see Costanza *et al.*, 1997b; Turner *et al.*, 2000). Therefore, here specific attention is devoted to this field. Environmental economics and ecological economics differ with regard to a number of theoretical and applied issues. A comparison can clarify this (for a more complete treatment, see van den Bergh, 2001).

As indicated in the previous section, the core of environmental economics is the theory of (negative) externalities or external costs. This considers environmental degradation and use of unpriced natural resources as negative effects by one economic agent on another outside the market, and without any form of compensation taking place. This implies that the environmental problem is cast in terms of an interaction between people (economic agents); that is, nature and environment are only implicitly described. Ecological economics is instead more interested in an explicit modelling of people–environment or ecological–economic relations, by mapping out cause–effect relationships and dynamic processes within the environment (hydrological, chemical, physical and ecological). According to Turner *et al.* (1997), this results from the fact that ecological economics is more closely related to traditional 'resource economics' and notably deals with renewable resources like fish, forests and water (Clark, 1990; Neher, 1990), rather than with environmental economics in a narrow sense ('economics of pollution').

Another important difference is between scale and allocation. Environmental economics aims for optimal allocation, and thus efficiency of use of scarce means (including resources). The objective is to find the optimal level of an externality, which follows from striving towards optimal social welfare or Pareto-efficiency. The latter is defined as a situation in which an improvement in the welfare of any individual cannot be achieved without a welfare loss for someone else. Environmental economics considers natural resources (gas, oil, fish, timber), environmental quality and services rendered by the environment and nature as scarce resources to which (optimal) allocation theories are applicable. Daly (1992) has long argued that economists have neglected the issue of an optimal physical scale or size of the economy and instead have focused completely on allocation issues. In the context of environmental sustainability and sustainable development goals, the scale problem has received much attention, shown also by academic and policy discussions about use of indicators to determine the physical dimensions of the economy (Gibson *et al.*, 2000).

Ecological economics has chosen sustainable development as its central concept. This is subsequently approached both qualitatively and empirically, with particular concern for equity, poverty and local or regional issues. Within environmental economics, sustainable development is usually regarded as being identical to sustainable growth at a national or global level (usually assuming a closed economy), which is studied with general and abstract models that avoid any reference to historical and spatial aspects, as well as specific characteristics of countries. Environmental economics does not seem to take absolute physical limits to growth as seriously as ecological economics and it regards the problem of a 'maximum scale' of the economy as irrelevant. A special point of attention in ecological economics is the structure and institutional context of developing countries. In addition, ecological economics generally assumes a longer time horizon than environmental economics and, consistent with this, pays more attention to cause–effect chains, interactions and feedback between natural and human–economic systems.

The concept 'co-evolution' is relevant here, as it is considered to reflect the mutual influence of internally diverse economic and environmental systems that creates a unique historical development. In this sense, ecological economics is closer in spirit to evolutionary than to neoclassical economics (van den Bergh and Gowdy, 2000). Evolutionary economics is characterised by concepts such as variety, selection, innovation, surprises, co-evolution, path dependence and lock-in. Path dependence and lock-in imply that possibly inferior technologies can become dominant as a result of unforeseeable historical events in combination with increasing returns to scale, caused by *inter alia*, positive network externalities (witness the market dominance of Microsoft operating systems). An implication of co-evolution and path dependence is that the market does not necessarily lead to a selection of (in a neoclassical sense) optimal technologies, production activities and use of space, even when prices are 'correct'. In line with this, ecological economics considers systems, including markets, as adaptive and coincidental rather than optimal.

Environmental and ecological economics also differ in their view on sustainable development. There are various definitions of sustainability and especially of sustainable development. Notably the differences between strong and weak sustainability have received much attention in the last few years (Ayres *et al.*, 2001). Weak sustainability has been defined on the basis of the concepts 'economic capital' and 'natural capital'. Economic capital comprises machines, land, labour and knowledge. Natural capital covers resources, environment and nature. Under weak sustainability, one strives to maintain total capital, defined as the sum of both types of capital. This allows the substitution of natural capital by economic capital, as has been analysed in economic growth theory. Strong sustainability, by contrast, requires that every type of capital is maintained separately. Environmental economics starts from weak sustainability, which emphasises a large degree of substitution of inputs in production and the economy as a whole. This has been criticised by proponents of ecological economics. Within ecological economics, usually some type of strong sustainability is emphasised, which is operationalised through goals such as the protection of critical ecosystems, striving towards a minimum-sized nature area or the maintenance of biodiversity.

The main criteria for evaluating developments, policies and projects differ between the two economic approaches. The dominant criterion of environmental economics is 'efficiency' (or sometimes a more limited version, such as cost effectiveness). Most economists would regard this as something trivial and hardly ethical. Nevertheless, it presumes that 'more is always better'. Furthermore, whereas in environmental economics distribution and equity are secondary criteria, ecological economics emphasises (basic) needs, 'north–south' welfare differences and the complex link between poverty and environment. In addition, ecological economics is best characterised by the 'precautionary principle', linked to environmental sustainability, with particular attention to 'small-probability–large-impact' combinations. This precautionary principle is closely related to a concern about the instability of ecosystems, the loss of biodiversity and environmental ethical considerations (bio/ecocentric ethics). Efficiency is of secondary concern in ecological economics. Distribution is often considered as a more important criterion for evaluating policies and changes than is efficiency. In addition, some argue that it is impossible to analyse distribution and efficiency separately. This would mean that the main tool of environmental economics, namely equilibrium analysis, which assumes that efficiency can be assessed independently of distribution, is inaccurate at best (Martinez-Alier and O'Connor, 1999).

Environmental economics focuses on value dimensions: utility and welfare in theory, and costs and benefits in practice. Unlike neoclassical economics, ecological economics does not regard a total valuation of (changes in) ecosystems as equal to the sum of private values. This is because the latter takes no account, or insufficient account, of internal environmental system functions, 'life-support' functions, values to future generations and non-instrumental existence values. Ecological economics is inclined to add criteria to the economic values for the purpose of correct decision making in the management of ecosystems. The term ecosystem health is used to cover aspects such as productivity, stability and resilience of ecosystems; biodiversity (genes, species, ecosystems); and the quality of the abiotic environment (Costanza *et al.*, 1992; see also the journal *Ecosystem Health*; and Sections 3.5 and 2.2.6).

Next, ecological economics criticises social objectives, such as those formulated within environmental economics, notably the utilitarian approach to intergenerational welfare. Alternatives are a Rawlsian principle of justice ('maximum criterion': Rawls, 1972) or a minimum welfare level encompassing (basic) needs (Stern, 1997). This is, of course, all just theory and can hardly be operationalised. In practice, the striving for income (gross domestic product: GDP) growth at a macro-level and cost–benefit analysis at a project level remain. So, which alternatives are offered by ecological economics in this respect? Some have pleaded in favour of physical or ecological indicators (material intensity per unit of service, 'ecological footprint'), which have as a main problem that they cannot accurately and unbiasedly describe systems at an aggregate level (van den Bergh and Verbruggen, 1999). Others have proposed a revised accounting framework to create a GDP type of progress or welfare indicator (e.g. Daly and Cobb, 1989). Still others are in favour of a multidimensional analysis based, for example, on multi-criteria evaluation (Martinez-Alier *et al.*, 1998; Munda *et al.*, 1994). In addition, the literature on ecological economics sometimes seems to discard consumer sovereignty when giving priority to the interest of systems above the freedom of choice of individuals, as is also done by environmental movements like 'deep ecology'.

Within ecological economics, a far-reaching integration is proposed between economics and ecology. Ecology is the area within biology that studies the relation of living organisms with their biotic and abiotic environment. It distinguishes various dynamic processes in ecosystems: population growth, ecosystem succession, shifting and multiple equilibria, and evolution. One operational technique for the aimed integration is – modelling at local, regional and global scales (see Braat and van Lierop, 1987; Costanza *et al.*, 1993, 1997b; van den Bergh, 1996). The wetland study presented in later chapters fits in this tradition and Chapter 3 offers a detailed exposition.

Proponents of ecological economics have often expressed dissatisfaction with the strict and fixed assumptions in traditional economic theory with regard to individual behaviour. These assumptions are usually summarised in the notion of 'unbounded rationality' and models of maximisation of profit (firms or entrepreneurs) and utility (households or consumers). These models underlie the analytical insights obtained by environmental economics with respect to economic valuation and environmental policy. Various branches of economics and closely related disciplines (such as evolutionary economics, institutional economics, experimental economics, psychology and sociology) have presented theoretically and empirically based critiques on these models. Although ecological economics seems to be sympathetic to these critiques, it has generated few alternative approaches so far. Van den Bergh *et al.* (2000) discuss the neoclassical approach, survey the criticism of it and present a first analysis of the implications of alternative models of individual behaviour for environmental policy. Such models include, among others, 'satisficing' (a combination of satisfying and optimising, i.e. bounded rationality), lexicographic preferences, relative welfare, habits and routines, imitation, reciprocal behaviour, changing and endogenous preferences and various models of behaviour under uncertainty. Spash and Hanley (1995) argue that lexicographic preferences offer an explanation for some of the problems met in economic valuation studies, notably that certain people are sometimes unwilling to make trade-offs between income compensation and environmental change (see also Blamey and Common, 1999).

Table 2.3 summarises the main differences between environmental and ecological economics. Note that this presents a somewhat simplified picture. Obviously, hybrid approaches are possible, especially as ecological economics is

Table 2.3 *Differences in emphasis between environmental and ecological economics*

Ecological economics	Traditional environmental and resource economics
Optimal scale	Optimal allocation and externalities
Priority to sustainability	Priority to efficiency
Needs fulfilled and equitable distribution	Optimal welfare or Pareto-efficiency
Sustainable development, globally and north–south	Sustainable growth in abstract models
Strong sustainability	Weak sustainability
Growth pessimism and difficult choices	Growth optimism and 'win–win' options
Unpredictable co-evolution	Deterministic optimisation of intertemporal welfare
Long-term focus	Short- to medium-term focus
Complete, integrative and descriptive	Partial, monodisciplinary and analytical
Concrete and specific	Abstract and general
Physical, biological and economic indicators	Monetary indicators
Systems analysis	External costs and economic valuation
Multidimensional evaluation	Cost–benefit analysis
Integrated models with cause–effect relationships	Applied general equilibrium models with external costs
Bounded individual rationality and uncertainty	Maximisation of utility or profit
Local communities	Global market and isolated individuals
Environmental ethics and intrinsic values	Utilitarianism and functionalism

diverse and not characterised by a rigorous theory. Moreover, some of the shortcomings of environmental economics (according to ecological economics) can be resolved within the traditional theoretical framework of neoclassical economics. For instance, environmental externalities can be modelled by describing dynamic causality relationships on the basis of ecological insights. Crocker and Tschirhart (1992) show that it is possible to incorporate descriptions of ecosystems within a wider framework of general equilibrium with externalities (see further in Section 2.4). Finally, the objective of sustainable development is also broadly supported nowadays by environmental economics, although definitions and interpretations are not always consistent with those adopted by ecological economics.

In evaluating the differences between ecological and environmental economics, it is interesting to note that one of the most influential biologists of the twentieth century, E. O. Wilson, has recently introduced the criterion 'consilience' as a measure of good science (Wilson, 1998). This denotes that the methods and starting points of one scientific discipline need to be consistent with the accepted insights of other disciplines, across all areas of science, including the natural and social sciences. Gowdy and Ferrer-i-Carbonell (1999) offer a discussion of consilience between biology and economics, the two most important disciplines supporting ecological economics, in order to examine to what extent economists and ecologists have influenced each other's way of thinking about environmental problems and their solutions. Within ecological economics, a dominant idea is that environmental economics approaches need to be made consistent with findings in ecology and thermodynamics. Nevertheless, this is complicated because of the distinct research traditions and methods of these two fields (Shogren and Nowell, 1992). In addition, consistency with insights from technical and other social sciences is also necessary. Insights from psychology and sociology could be useful, particularly with regard to modelling consumer and company behaviour for environmental policy analysis and monetary valuation.

2.4 Natural–social science integration

This section presents available frameworks and conceptual perspectives underlying the integration of economics, ecology and other disciplines. The literature shows a wide variety of such frameworks. Surveys are offered by Barbier (1990), van den Bergh and Nijkamp (1991), van den Bergh (1996), Costanza *et al.* (1997b), Ayres *et al.* (1999) and Turner *et al.* (1999). A very general and almost non-theoretical (few assumptions or restrictions) framework is the driver–pressure–state–impact–response (DPSIR) framework, a variation on the pressure–state–impact–response (PSIR) framework proposed by the Organization for Economic Cooperation and Development (OECD, 1995) for environmental data classification, proposed by Rotmans and de Vries (1997) and Turner *et al.* (1998b, 2000) as a conceptual basis for integrated analysis and modelling. The five components of the framework have the following interpretation:

driver: economic and social activities and processes;

pressure: pressures on the human (health) and environmental system (resources and ecosystems);

state: the physical, chemical and biological changes in the biosphere, human population, resources and artefacts (buildings, infrastructure, machines);

impact: the social, economic and ecological impacts of natural or human-induced changes in the biosphere; and

response: human interventions on the level of drivers (prevention, changing behaviour), pressures (mitigation), states (relocation) or impacts (restoration, health care).

According to Rotmans and de Vries (1997), there are various types of integration. *Vertical integration* means that the causal chain in the PSIR or DPSIR framework is completely described in a model ('close the PSIR loop'). *Horizontal integration* (of subsystems) in this context is defined as the coupling of various global biogeochemical cycles and earth system compartments (atmosphere, terrestrial biosphere, hydrosphere, lithosphere and cryosphere). *Full* or *total integration* means a combination, leading to the complex linking of various drivers, pressures, states, impacts and responses, thus allowing for various synergies and feedbacks.

Most other integration frameworks proposed in environmental and ecological economics represent more specific theoretical choices than the DPSIR model. We discuss several below.

Crocker and Tschirhart (1992) formulate a static general equilibrium model with externalities occurring via ecosystem food chains (see also Crocker, 1995). This allows for tracing indirect and complex processes linked to externalities and, connected with this, to policy rules that take such complexity well into account. The authors even go one step further by modelling ecological relationships as resulting from decisions made by organisms trying to optimise stored energy in their biomass. General equilibrium systems are generalised input–output systems, which include non-linear processes (substitution), endogenous prices (allocation and demand–supply interaction) and behavioural objectives (profit and utility maximisation). Usually these models are static, as the combination of general equilibrium and dynamics creates very complicated conceptual and analytical problems (relating to perfect foresight and intertemporal optimisation). The simpler input–output system was, in fact, proposed as one of the first conceptual models of linked ecological–economic systems (Daly, 1968; Cumberland, 1966; Isard, 1969, 1972; Perrings, 1987). Input–output models and models with price mechanisms have also been applied to ecosystem processes (see Hannon, 1973, 1976, 1986, 1991).

Siebert (1982) considers a very simple but complete dynamic representation of the combined effects of resource use and pollution linking an economy to a renewable resource. This idea has been generalised through the concept 'environmental utilisation space' (Opschoor and Weterings, 1994), which denotes the larger system that is essential for the functioning of a renewable resource (say a fishery). This larger system includes its life-support (eco)system and the wider economic, social and institutional context that influences the resource via markets, regulation, culture, economic growth and technical change. Van den Bergh (1993) proposes a generalisation of the idea of mutual impacts of economic and ecological systems, which gives rise to a model with continuous feedback loops. A simple dynamic system consisting of two equations can be regarded as a conceptual framework for the integration of economic and environmental systems or processes. Such a system is able to generate the universal growth–feedback–collapse patterns. Although ecologists will regard the basic logistic growth representation of a renewable resource as too simple to be useful in applied studies of ecosystems, many economists make ample use of it in both theoretical and applied research (see Wilen, 1985). The notion of renewable resources, subject to logistic growth and now and then to more complex processes, is common in the literature on fisheries, water and forestry economics (see Aronsson and Löfgren, 1999; Clark, 1990; Eggert, 1998; Zilberman and Lipper, 1999). Swallow (1994) integrates theoretical models of renewable and non-renewable resources to address multiple use and trade-offs in wetland systems.

While resource models are usually partial in economic scope, because of their focus on resource management and sometimes product markets, more general models have addressed wider issues of economic growth and development, recently in the context of sustainable development. A range of approaches can be identified (see van den Bergh and Hofkes, 1998). At a theoretical level, exogenous and endogenous economic growth theories with environmental variables can be regarded as examples of integrated modelling. They link a description of economic development – in terms of investments, population growth and technological change – with simple dynamic descriptions of natural processes (Gradus and Smulders, 1993; Kamien and Schwartz, 1982; Tahvonen and Kuuluvainen, 1993). The modern treatment of growth and environment at an aggregate level is most clearly reflected by the economic growth theory approach to (weak) environmental sustainability (see Toman *et al.*, 1995) and the macroeconomic models of climate change (Nordhaus, 1994; for a concise survey, see van Ierland, 1999).

Holling (1986) offers lessons for ecosystem management based on a 'four box model' that depicts ecosystems as following a cyclical pattern of exploitation, conservation, release (creative destruction) and reorganisation. Batabyal (1999) has translated this idea into a simple formal framework to study optimal management of cyclical ecosystems. In another publication, Holling (1978) has laid the foundation for 'adaptive management'. This is based on the idea that complex, uncertain and uncontrollable systems like ecosystems and economies require that experimental research, careful monitoring, policy learning and adaptive management are combined (Gunderson *et al.*, 1995; Lee, 1993; Walters, 1986). Among other things, this line of thought has resulted in multiagent dynamic models at ecosystem and global scales, with agents behaving as satisficers (a combination of satisfying and optimising) that adapt to changes in their environment (Janssen, 1998a,b; Janssen and de Vries, 1998; Janssen *et al.*, 1999). This can be considered as an integration of evolutionary and ecological economics. Finally, Holling (1998) compares analytical and integrative cultures in ecology and argues that both are needed, and need to interact, in order to avoid falling into the trap of providing precise answers to wrong questions or useless answers to the right questions. Moreover, the integrative culture can connect social science and policy analysis.

The previous approach provides one elaboration or interpretation of resilience. This has been defined in various ways (Perrings, 1998): as the ability of systems to maintain structure and functions in the presence of external stress; as the magnitude of disturbance that can be absorbed before a system flips from one state to another (Holling, 1973); and as the time taken to return to a steady state following a perturbation (Pimm, 1984). Common and Perrings (1992) argue that resilience is the essence of the ecological approach to sustainability and that it implies an entirely different approach to the one used by mainstream economists, which is based on aggregate growth theory. Levin *et al.* (1998) suggest that resilience is an enlightening concept for the study of both ecological and economic systems. Various invited responses to this article in the same journal (*Environment and Development Economics*, 3, 221–262), however, criticise this point of view, stating that resilience offers nothing new and can be neither measured nor quantified.

Recently, biodiversity has received much attention, mainly at a conceptual–theoretical level (Perrings and Pearce, 1994; Perrings *et al.*, 1995a,b; Weitzman, 1998). The relationship between biodiversity and resilience is a focal point in this context (Holling *et al.*, 1995). One view on this relationship says that species have overlapping roles and, therefore, adding or losing species causes little or no change until a threshold is reached. Another view is that species in an ecosystem can be divided into functional groups of species: 'drivers or keystone species', and 'passenger species'. The latter may seem redundant but can take the role of keystone species as environmental conditions change, thus ensuring resilience through their ability to adapt to new conditions. The modelling of these issues has just started (Perrings and Stern, 2000; Perrings and Walker, 1997). Weitzman (1998) presents a cost-effectiveness model for deciding about biodiversity protection options, in particular preservation actions that improve the survival probability of a particular species. This model distinguishes between direct utilitarian values and indirect values represented by a 'distinctiveness' indicator. The latter is formalised via the expected (genetic) distance between the genes of a particular species and the genes of the closest resembling species. The more classical way of modelling species extinction originates from fisheries economics, which shows that lack of property rights and high discount rates relative to rates of return to the fishery activity are critical parameters in determining whether a fishery will become extinct (see Clark, 1990). Swanson (1994) has applied a similar model to land-based resources. This model makes it possible to address issues such as alternative land uses and associated rates of return, as well as the influence of population pressure, development and management choices.

The physical dimension of economies and economy–environment interactions has been explicitly addressed by applying the two main laws of thermodynamics. The classic publications are Boulding (1966), Daly (1968), Ayres and Kneese (1969) and Georgescu-Roegen (1971). Ruth (1993) offers an ambitious effort to integrate these physics-oriented perspectives with the ecology-based perspectives in integrated modelling. Van den Bergh and Nijkamp (1994a) is an example of the integration of ecological and economic systems in a dynamic context through the explicit description of physical processes and flows in and between these systems.

A final class of conceptual approaches pays attention to evolutionary aspects in resource dynamics, in addition to the so-called 'mechanistic' (i.e. repeated, predictable and reversible) dynamics, such as regeneration, assimilation and accumulation of capital (growth) and pollution (Gowdy, 1999; van den Bergh and Gowdy, 2000). Although these approaches are diverse, they seem to agree that changes in economic reality arise so quickly that it is impossible to maintain or even reach an equilibrium state. An example of an evolutionary process is the selection impact of pesticide use on the proportion of genetically resistant individuals in a pest species over time, which creates negative externalities for the future (Munro, 1997; Norgaard, 1994). A similar process occurs in fisheries, as fishing-net mesh size generates selective pressure on fish size so that relatively small fish can increase their proportion in the fished population (Allen and McGlade, 1987). At a higher level, Wilkinson (1973) argues that ecological factors played a crucial role in the conception of the Industrial Revolution as forests became depleted causing scarcity of wood, in response to which investments were made in alternative energy sources (notably coal); this, in turn, created stimuli for the application and further development of certain technologies (notably the steam engine) (see also Common, 1988; Faber and Proops, 1990). A general lesson stemming from evolutionary biology is that seemingly redundant features of ecosystems, economies and even human cultures should be preserved and that maximum attainable diversity should be an important goal (van den Bergh and Gowdy, 2000). This is necessary to sustain 'evolutionary potential' and to maximise adaptive flexibility of future economic and environmental systems. This conclusion is equally consistent with the ideas of resilience, 'precautionary principle' and 'safe minimum standard' (Ciriacy-Wantrup, 1952; Crowards, 1998).

2.5 Conclusions

This chapter has offered a wide perspective on elements of the natural and social sciences that are, or can be, used in the study of wetland ecosystems. Given the multifaceted nature of wetlands, their processes, functions, values and problems, a multidisciplinary perspective is required to study them thoroughly.

With the concepts and theories presented here, the reader will be able to understand most of the subsequent discussions, models and analyses. Whereas the last section on integration has adopted, purposefully, a conceptual–theoretical perspective on integrated research involving certain natural and social sciences, the next chapter will move the discussion onto a more operational level of integration, by presenting a survey of issues relating to integrated modelling and assessment. Indeed, integrated modelling can be seen as the best, or even the only, approach to integrate systematically the wide range of hydrological, ecological and economic aspects of wetlands.

3

Integrated modelling and assessment

3.1 Introduction

Integrated ecological–economic modelling has been practised since at least the late 1960s. Russell (1995) presents an interesting *ex post* evaluation of the early work undertaken at Resources for the Future. He is modestly optimistic about the feasibility of the formal linking of economic and ecological models but argues that significant financial and human resource investments are required to improve the quality of integrated modelling. Such investments have been undertaken in some areas of application, notably relating to the enhanced greenhouse effect and climate change, but are less evident in the area of ecosystem modelling. The latter has seen mainly *ad hoc* efforts at integrated modelling, and certainly less interaction and competition among different research teams.

The types of question addressed by integrated models have changed over time. The early exercises were oriented towards the spatial dispersal of discharges of pollutants to air and water and the subsequent impacts on environmental quality and ecosystems. Later efforts have been aimed at a range of issues such as the modelling and testing of:

- the stability and resilience of complex systems (Holling, 1978; Walters, 1986);
- the impact of multiple stress factors (Braat and van Lierop, 1987);
- the impact of simultaneous ecological and economic shocks (e.g. floods and droughts) (Moreira *et al.*, 2001; Rietkerk *et al.*, 2000);
- the relationship between development, resources and environment (sustainable development) (Arntzen, 1989);
- the impact of nature policy and management (Chapin *et al.*, 2000; Pimm and Lawton, 1998);

- the influence of land use on environmental quality and landscape (this study);
- the influence of hydrological management in wetlands (Maltby *et al.*, 1996);
- the impact of acid rain on vegetation and lakes (Alcamo *et al.*, 1990); and
- the impact of global environmental processes (deforestation, biogeochemical cycles (various IPCC (Intergovernmental Panel on Climate Change) studies).

Obtaining an overview of the entire field of integrated modelling is difficult if not impossible. Various traditions exist, sometimes indicated by different terminology (for instance, 'integrated modelling' versus 'integrated assessment'; see Rotmans and Vellinga, 1998). Many models are too complex to be presented in journals or remain simply unpublished. The heterogeneity of integrated modelling creates a serious problem for the transferability of models and results to different contexts, regions or periods. Unavoidably then, the wheel is continuously reinvented. Moreover, modelling ambitions and resources or capacities remain unbalanced unless teams of experts from different disciplines are brought together, at various levels. State research institutes like the RIVM (National Research Institute for Public Health and Environment) in the Netherlands have produced the IMAGE and TARGETS models (see Alcamo, 1994; Rotmans and de Vries, 1997). At the international level, a panel like the IPCC has created a dialogue between natural and social scientists as well as within each of these groups. The case study presented later in this book is based on a joint project in which various (multidisciplinary) university institutes collaborated, bringing together economists, ecologists and hydrologists.

Formal modelling and evaluation in integrated economic–environmental studies has both advantages and pitfalls. Its main advantages are handling, in a systematic and consistent way, data, information, theories, and empirical findings from various contributing disciplines; being explicit about assumptions, theories and facts; and, addressing complex phenomena, interactions, feedback, laborious calculations and temporally, spatially and sectorally detailed and disaggregated processes. Pitfalls include unclear synergy of approximations and uncertainties; rough application of monodisciplinary theories and empirical insights; simplification of complex phenomena (e.g. by treating them as a black box); misinterpretation and arbitrary choice of disciplinary perspectives by the modeller; and lack of systematic or complete linking of subsystems or submodels. Complex or large dimensional models have the extra disadvantages of being difficult to calibrate and validate and of lacking transparency.

An argument against non-formal approaches to integrated research is that these fail to provide a systematic and consistent linking of data, theories and empirical insights from various disciplines. Instead, these approaches tend to result in a battle of perspectives based on distinct and usually implicit premises and information bases. Models force researchers at least to be explicit about the latter two inputs to integrated research.

Perhaps the main disadvantage of models is that they are trusted too much; that is, they run the risk of being interpreted as objective representations of reality and are then taken too seriously especially by laypersons and policy makers. Yet, policy makers also often indicate their doubts about formal modelling. Shackley (1997) states that numerical models have, despite their long tradition of development and widespread use, not achieved the epistemological status that the controlled laboratory experiment has in natural sciences (and more recently in social sciences and environmental economics in particular; see Shogren and Hurley, 1999). This relates to the fact that modelling results never 'prove' anything, since they do not generate real or physical processes. The best way to view theoretical and especially empirical models is to consider them as tools for hypothetical experiments with complex systems, which serve as analogies or pictures of real-world systems that do not allow – technically, morally or politically – for experimentation. In other words, complex model systems, notably integrated ecological–economic models, are heuristic devices for learning about the real-world system, rather than for predicting its real course. This may not work, if only because model predictions can themselves influence political decisions – in economics known as the Lucas critique (Sargent and Wallace, 1976).

This chapter provides a survey of frameworks and methods of integrated ecological–economic modelling. Section 3.2 discusses general aspects of integrated modelling and assessment. Section 3.3 concentrates on spatial aspects of integrated modelling, as these are central to the present study. Section 3.4 discusses the potential relationships between integrated modelling and monetary valuation. Section 3.5 provides a background for the construction of environmental quality – sustainability and ecosystem health – indicators. Section 3.6 discusses evaluation issues, in particular related to the use of multicriteria techniques in evaluating multidimensional output generated by integrated models.

3.2 Integrated modelling

In order to address questions about sustainable development – at the ecosystem, local, regional or global scale – models can be used that integrate economic and ecological modules. Such integrated models are usually employed for policy or scenario analysis, and to a lesser degree for the purposes of policy design, scenario generation and forecasting. A general method used to develop integrated models is a systems approach (also known as systems dynamics). This covers a wide range of model types: linear versus non-linear, continuous versus discrete, deterministic versus stochastic, and optimising versus descriptive. The systems approach makes it possible to deal with concepts, such as dynamic processes or feedback mechanisms, and control strategies (see Bennet and Chorley, 1978; Costanza *et al.*, 1993). One can integrate two subsystems or have a hierarchy or nesting of systems. The fixed elements in the system can be either considered as black boxes or described as empirical or logical processes themselves. The systems approach is suitable for integrating existing models and for incorporating temporal as well as spatial processes. Costanza *et al.* (1993) classify modelling approaches according to whether they optimise:

- generality: characterised by simple theoretical or conceptual models that aggregate, oversimplify and exaggerate;
- precision: characterised by statistical, short-term, partial, static or linear models with one element elaborated in much detail; or
- realism: characterised by causal, non-linear, dynamic–evolutionary and complex models.

These three criteria are usually conflicting, so that a trade-off between them is inevitable. A distinction is relevant here between analytical and heuristic integration. Analytical integration means combining all aspects studied in a single model (and, therefore, a single model type). Heuristic integration denotes using the output of one model as an input to another, and vice versa, and possibly extending this by a (finite) iterative interaction. In this case, different model types can be combined, such as optimisation with descriptive. If one wants to attain a great deal of analytical power then analytical integration seems attractive, whereas striving for realism might imply the use of a heuristically linked set of models that each deal with a particular environmental dimension, problem or compartment. The development of integrated models, by the joint effort of economists, ecologists and others, is based on bringing together elements, theories or models from each discipline and transforming them for the purpose of integration. This may require steps like reduction, simplifying or summarising. The results may not always be greeted with enthusiasm within the related discipline, as certain nuances will get lost through the various steps.

The remainder of this section offers a selective review of approaches to, and applications of, integrated modelling. This will allow the reader to get a sense of the variety of approaches. Many integrated models defined at the level of ecosystems are based on the standard systems–ecological approach (Jørgenson, 1992; Patten, 1971). They include ecosystem modules that describe the effects of environmental pollution, resource use and other types of disturbance. A main problem is modelling the effects of multiple synergetic stress factors, as the empirical basis for this is often lacking. Various integrated models have been developed for terrestrial and aquatic systems. Surveys are presented in Braat and van Lierop (1987), van den Bergh (1996) and Costanza *et al.* (1997b). Some studies have paid considerable attention to spatial aspects, focusing on spatial disaggregation into zones (e.g. Giaoutzi and Nijkamp, 1993; van den Bergh and Nijkamp, 1994b) or land-use planning in interaction with landscape ecology (see Bockstael *et al.*, 1995; Section 3.3). Formal theoretical approaches in ecology that provide a basis for these approaches have been described by Watt (1968), Maynard Smith (1974), Roughgarden *et al.* (1989) and Jørgenson (1992). Perrings and Walker (1997) consider resilience in a simple integrated model of fire occurrences in semi-arid rangelands, such as are found in Australia. The model describes the interaction between extreme events (fire, flood and droughts), grazing pressure and multiple locally stable states. Carpenter *et al.* (1999) develop and explore water and land-use options in an integrated model of a prototypical region with a lake that is being polluted. This model combines behaviour, which exhibits bounded rationality supposedly in accordance with the reality of regional resource and environmental management, and a non-linear ecosystem module describing processes occurring at different speeds. The model generates multiple locally stable states as well as 'flipping' behaviour (see also Janssen *et al.*, 1999).

A special category of integrated modelling is sometimes referred to as the biophysical or energy approach. This aims to integrate economic, environmental and ecological processes in energy–physical dimensions, based on the notion that any system is constrained by energy availability (Odum, 1983). These models include energy and mass balances. A central concept in this approach is 'embodied energy', defined as the direct and indirect energy required to produce organised material structures. Applications of these energy-inspired models occur for ecosystems, economic systems and economy–environment interactions (Odum, 1987). An extended application to a regional system is given by Jansson and Zucchetto (1978; see also Zucchetto and Jansson, 1985) and a concise survey of this approach and its applications is given by Herendeen (1999). A recent overview and application is given by Hall *et al.* (2000).

Integrated models have also been applied at a global scale, beginning with the model presented in the first report to the Club of Rome (*The Limits to Growth*; Meadows *et al.*, 1972). Meadows *et al.* (1992) provides an updated version of the original model with similarly pessimistic conclusions. Meadows *et al.* (1982) offer an overview of the most important 'world models' of the 1970s by various authors and modelling teams, which shows an evolution towards increasing detail and disaggregation of sectors and space. The original model in *The Limits to Growth* was heavily criticised by economists and econometricians. The first group accused the modelling team of neglecting to incorporate market processes (prices), which would generate substitution and technical change as solutions to global resource scarcity and environmental pollution problems. The second group argued that a systems dynamics model has a weak empirical basis, because the model structure is not subject to rigorous statistical estimation and no statistical performance indicators and validity tests can be applied to the model as a whole. Econometric estimation of systems of equations would provide an alternative. In defence, one can say that long-term projections, as undertaken with global or world models, do not gain much from adding prices and using sophisticated statistical techniques to estimate model parameters. The reason is that historical patterns cannot be assumed to extrapolate in a straightforward way into the future. This explains why various approaches can presently be found in integrated modelling; the main categories are econometric modelling, input–output modelling, systems dynamics modelling, computable (partial or general) equilibrium modelling, optimisation (linear, non-linear, integer, mixed and dynamic programming) and a mix of these pure modelling types.

The recent focus on integrated assessment of the enhanced greenhouse effect (potential climate change) can be regarded as the new wave in world models, where (again) economists and others have tended to rely on different model approaches (Bruce *et al.*, 1995). The integrated climate assessment models provide integration of natural sciences (physics, chemistry, biology, earth sciences) and social sciences (economics, sociology, political science) and have so far given rise to a continuation of the trend in world models towards increasing detail and disaggregation. These climate assessment models often have a multilayered conceptual structure that distinguishes physical and environmental effects of human activities from adjustments by humans to climate change (individuals, firms, organisations) and policy responses (mitigation, aimed at the causes) at various spatial levels (see Parry and Carter, 1998).

Integrated models can have different formats. One important distinction is between policy optimisation and evaluation (usually numerical simulation) models. One of the first, and famous, integrated assessment models used in policy making is the RAINS model (Alcamo *et al.*, 1990). This includes an optimisation algorithm for calculating cost-effective acidification strategies in Europe, aimed at realising deposition targets throughout Europe and taking account of sensitive natural areas (forests and lakes). This model is a rare case of direct science–policy influence, as it was used in the negotiations on transboundary air pollution in Europe. Castells Cabré (1999) offers an informative analysis of the institutional and evolutionary dimensions of the interaction between scientists, research institutions and negotiations on international environmental agreements, with special attention given to the RAINS model and the acid rain context in Europe.

In the area of integrated assessment models for carbon dioxide emission (climate) strategies, one can find both economic optimisation (Nordhaus, 1994) and detailed descriptive model systems such as IMAGE and TARGETS (Alcamo, 1994; Rotmans and de Vries, 1997). DICE by Nordhaus (1991) is the first example of a policy optimisation model for climate change. The model essentially combines economic growth theory with a highly aggregated climate change model. Tol (1998) provides a short account of the evolution of the economic optimisation approach to climate change research. He emphasises the attention given to the analysis of uncertainty and learning from a cost-effectiveness perspective, which has given rise to various model formulations and analyses.

CETA and MERGE are, like DICE, policy optimisation models, in the cost–benefit or economic growth tradition, that include extremely simplified versions of biophysical models (see van Ierland, 1999). IMAGE and TARGETS fall into the systems dynamics approach. Some other models follow a stochastic simulation approach (see Janssen, 1998b). The models ISLAND (Engelen *et al.*, 1995), Quest (Biggs *et al.*, 1996) and Threshold (Millennium Institute, 1996) have been developed for regional scale analysis. Janssen (1998a,b) mentions complex adaptive systems as a separate category that differs from complex systems (the basis of the systems dynamics approach) in that it assumes no equilibrium (neither single nor multiple) and is characterised by multiple heterogeneous agents, unpredictability, irreversibility and adaptive and evolutionary mechanisms.

Janssen (1998b) and van Ierland (1999) present informative surveys and categorisations of macroeconomic–environment and macrolevel integrated models, including the climate-oriented integrated assessment models. Van Ierland devotes special attention to the various ‘regionalised world models’ (with acronyms like RICE, CETA, MERGE, DIALOGUE, FUND). Van den Bergh and Hofkes (1998) collect distinct approaches to integrated models with an economic emphasis focusing on sustainable development questions, in theory and practice as well as at global and regional levels.

Many applied integrated models have a disaggregated structure. This allows the examination of a sustainable economy in terms of its sectoral composition of output. The main approach in this respect is the input–output model, which describes sectoral interactions as linear relationships based on fixed input–output production coefficients. This approach allows the calculation of all indirect sectoral, resource-use and emission effects of different final demand compositions to be calculated. Duchin and Lange (1994) undertook an important study using the input–output approach. They extended the Leontief world model (Leontief *et al.*, 1977) to test the Brundtland Commission's statement that growth and sustainability go well together (WCED, 1987). Their conclusion was that the statement did not hold true. The model they used describes the period 1980 to 2020 and covers 16 regions and 50 sectors, as well as dynamics in the trade of commodities, flows of capital and economic aid. It calculates the use of energy and materials (including metals, cement, pulp, paper, chemicals), emissions of carbon dioxide (more than doubled worldwide over the studied period), sulphur dioxide (almost constant) and nitrogen oxides (almost doubled).

Another type of multisectoral approach is formed by computable general equilibrium (CGE) models. These describe interactions between sectors as either linear (input–output structure) or non-linear. Computable general equilibrium models are particularly useful when one wants to study the economic structure effects of environmental policies. Nevertheless, dynamic formulations can become extremely complex (and require strict assumptions). They do not seem very suitable for integration with environmental or ecological processes, even if purely theoretical models have provided some basis for this (see the discussion of the approach by Crocker and Tschirhart (1992) in Section 2.4).

The new trend in integrated modelling and analysis seems to consist of adaptive management and participatory approaches (Gunderson *et al.*, 1995; Gunderson and Holling, 2001; Holling, 1978; Lee, 1993). This is based on the idea that experts do not agree on the best model and theory, particularly in the context of complex systems and the questions that have stimulated the use of integrated modelling. Another feature of such problems reinforces the relevance of adaptive management: namely, the presence of stakeholders with sharply opposite interests and value systems. The implication seems to be that modelling should be turned into a democratic exercise. The 'post-normal science' of Funtowicz and Ravetz (1994) offers a similar perspective, arguing that, in situations where decision stakes and systems uncertainty are high, the peer community needs to be extended with investigative journalists and laypeople, among others. Only time will tell whether these ideas are merely fashionable or a resilient feature of integrated modelling. Table 3.1 illustrates the characteristics of integrated models and provides some general examples.

3.3 Spatial modelling

An important problem in integrated modelling is the linking of processes or systems at various scales, both within and between different disciplines. Causes of environmental pressure can be local (fragmentation, desiccation), regional (acid rain, river pollution) or global (global warming). Ecology and hydrology usually employ very strict boundaries and clearly describe local cause–effect chains. Economics often uses less-clear spatial boundaries and describes external factors without clearly specifying their spatial sources. This difference is related to the fact that the extent (i.e. the size) of the problem, in terms of spatial, temporal and conceptual dimensions, is better bounded and often smaller in the

Table 3.1 *Characterising integrated models*

Model criterion	Range of choice	Examples of distinct approaches
Analytical integration	Optimisation (benevolent decision maker); equilibrium (partial or general); game–theoretical; dynamic–mechanistic; adaptive (multiagent and dynamic); evolutionary (irreversible, bounded rationality)	Many theoretical models: growth theory; renewable resource economics (fisheries, forestry, water quality/quantity); systems models (limits-to-growth; Meadows *et al.*, 1972); cost-effectiveness models (RAINS); welfare optimisation (DICE)
Heuristic integration	Satellite principle; multilayer subsystems; sequential; parallel consistent scenarios; aggregation of indicators; evaluation	Regional environmental quality models (Resources for the Future); World models (Club of Rome); integrated assessment models; the present study
Spatial coverage	World; national; regional; urban, local, ecosystem	Ecosystem modelling; macroeconomic modelling; regional modelling; urban modelling; World models
Spatial disaggregation	Single region; multiregion; spatial grid (geographical information systems)	Integrated assessment models (climate change); land-use models
Mathematical structure	Static equilibrium (non-linear system of equations); static input–output (linear system of equations); static (linear or non-linear) optimisation; dynamic simulation (stock-flow); dynamic input–output; optimal control; dynamic programming; multicriteria/multiobjective	DICE, RAINS (optimistic); IMAGE, TARGETS (numerial simulation)
Disciplines involved	Demography; microeconomics (equilibrium theory); macroeconomics (growth theory); regional economics; population ecology; ecosystem modelling; hydrology; physical geography; social geography; sociology; political science; decision science	Gilbert and Braat (1991; demography); DICE (growth theory)

Table 3.1 (cont.)

Model criterion	Range of choice	Examples of distinct approaches
Aggregation level	Micro (individuals, households); macro (national economy, main sectors, global); sectoral; interest groups; homogeneous land plots; spatial grids; temporal (days, seasons, years)	Computable general equilibrium models; macroeconomic models (Keynesian); multisector models; land-use models; landscape models

natural than in the social sciences. The advantage of a smaller extent is that it allows for a higher resolution (i.e. precision in measurement and description). A larger extent is usually accompanied by lower resolution and more aggregation of basic units or processes.

Gibson *et al.* (2000) discuss the issues surrounding scaling in integrated modelling, defining scale as 'the spatial, temporal, quantitative, or analytical dimensions used to measure and study any phenomenon' (p. 218). A relevant subdivision of scale is into absolute and relative scales. The first is objectively measurable and fixed, whereas the second depends on the context, often related to the movement of species or humans, in which speed (or time) are important and variable parameters. 'Level' is a concept related to scale; level is defined as 'the units of analysis that are located at the same position on a scale'. Different levels can be linked in a hierarchy, of which there are various types. The most relevant for formal modelling seems to be the constitutive hierarchy, in which groups of objects or processes are combined into new units, which results in specific functions at each different level. These are called 'emergent properties'; that is, properties that cannot be derived from knowledge of characteristics of units and processes at a lower level. This issue has been much debated in evolutionary biology and economics; in the latter, it is known as the microfoundations debate (see van den Bergh and Gowdy, 2003). A large number of levels can be identified (or better, defined) when going from very micro (individual organism) to very macro (the biosphere or earth) scale. In biology this has led to various subdisciplines, each concerned with a particular level of study (molecular biology, cell biology, histology, physiology, population ecology, ecosystem ecology and biogeography). In economics, however, microeconomics has come to dominate all the levels of study, such as technological innovation, urban economics, regional economics, public economics, macroeconomics and international economics.

A hierarchy of levels implies that, at each level, processes or units can be described in terms of influences from units at the same level or from units at higher (downward causation) or lower (upward causation) levels. In economics, these latter approaches are sometimes referred to as top-down and bottom-up. Gibson *et al.* (2000) mention in this context 'hierarchy theory', according to which a process at a certain level is bounded by constraints provided by higher and lower levels. An alternative view is to distinguish between processes at different levels occurring at different speeds (slow- and fast-moving variables). Attractive as these ideas may seem, various conceptual and practical problems surrounding this approach have not yet been satisfactorily resolved.

Costanza *et al.* (1997a, p. xxii) state that the integration of economics and ecology is hampered by the lack of space in economic theories and models. Although it is true that mainstream economics has largely assumed away space and spatial externalities between economic agents, the statement neglects the large area of spatial economics. This covers regional, urban and transport economics as well as spatial informatics – mainly the application of geographical information systems (GIS; see Scholten and Stillwell, 1990). Various linkages exist between these research areas and environmental economics, focusing on location choice, diffuse pollution, land use and transport (see the various surveys in van den Bergh (1999), Part V).

GIS applications are nowadays often considered as an essential input to integrated spatial models, because they allow for the interaction between economic and ecological phenomena to be represented at a detailed spatial scale. However, it is not always clear beforehand that using a high spatial resolution will be fruitful. Whereas many ecological and hydrological processes are suitable to a grid-based description, most economic processes operate at higher scales. This explains, for instance, why a method like 'cellular automata' has been more popular in landscape ecology than in spatial economics (Engelen *et al.*, 1995).

The cellular automata approach in a pure form describes discrete time transformations of discrete cell states in a regular n-dimensional space (usually $n = 2$), where the transformation is completely determined by the states of the cell itself and of its immediately neighbouring cells (a generalization of the well-known 'game of life'). Given the initial states of all cells, a spontaneous spatial evolution will take place. Cellular automata were originally used to model process-based predator–prey interactions, changes in surface water quality, and fire propagation problems. They allow for an explicitly spatial process approach. Whereas such immediate influences in space dominate in physical and biological systems, this is not the case in social and economic systems. Here 'spaceless' information is usually crucial to individual and public decisions. Moreover, many spatial interactions in an economic context extend directly beyond the scope of 'neighbouring cells': the extreme examples being international trade and globalisation.

Nevertheless, perhaps the cellular automata approach can address issues in social science that depend on the physical and network interactions between people in local neighbourhoods. In this sense, it would act as another example of a micro- or bottom-up model that can generate macropatterns. The cellular automata method has already been applied to urban growth, infrastructure development and forest clearing. Describing land-use patterns with cellular automata is particularly relevant when physical planning and land regulation are lacking or lax, as spatial processes are then unrestricted and 'spontaneous'. An example of an integrated model with cellular automata for analysing land-use patterns is given by White and Engelen (1997). This model combines various types of suitability (environmental, institutional, neighbourhood) and accessibility indicator to trace endogenous land-use patterns in the context of a regional economic–environmental system. Crucial to such an application is the fact that institutional restrictions are not so strict that they completely dominate land-use patterns, as is the case nowadays in many Western countries, notably the densely populated ones such as the Netherlands. In this case, there is no room for spontaneous spatial evolution. The application of cellular automata in social science, therefore, seems very limited, unless generalisations and extensions are used – and further developed – that allow for direct interactions between non-bordering cells (see Couclelis, 1985).

Within economics, urban and regional economics have provided space-oriented inputs to integrated modelling (Brouwer, 1987; Hafkamp, 1984; Nijkamp, 1979a,b). These usually involve regionally disaggregated models that pay attention to regional location; interregional flows of products, capital and labour (migration); interregional externalities; or hierarchical planning problems. At a multiregional level, a distinction is usually made between bottom-up and top-down models, which, confusingly, have been defined in different ways by different authors. Bottom-up models can be defined as adopting a starting point at the level of individual agents or regions, and allowing for feedback from the lowest to the highest (macro) level. Top-down models are steered or dominated by relationships among aggregate variables and ultimately by exogenous macrolevel developments. Multiregional models following the bottom-up format are the most data demanding. For more discussion of regional and multiregional aspects in integrated models, see Nijkamp *et al.* (1986) and Lakshmanan and Bolton (1986).

Landscape ecology, originating from an interaction between ecology, geography and land-use planning, has developed into a field that can provide another important spatial information input to integrated modelling. It studies the two-sided link between ecological processes and spatial patterns in the environment, taking account of the spatial heterogeneity of land cover (i.e. vegetation and habitat types). The spatial patterns of nature and environment change over time through various mechanisms (Turner, 1998): (i) regular and short-term natural processes (e.g. ecosystem succession, sedimentation and erosion) and less-regular causes (e.g. fires, pests and storms); (ii) long-term evolution caused by local systems developing in different directions as a result of being spatially isolated for long periods of time; and (iii) land use by human activities. The last has especially caused fragmentation of nature, through land clearing, land use, settlements and infrastructure. An important question is whether the spatial structure or mosaic (i.e. the spatial organisation of land cover viewed from above) will after some time move to a steady state or be continuously changing. Local and regional climate, topography, soil types and hydrological features form the boundary conditions for the range of feasible land-cover patterns. The amount of spatial detail in landscape ecology studies partly determines how spatial interactions will be perceived, so choice of the level of spatial detail can have a significant influence on the outcome of an analysis. Important concepts in landscape ecology are function (i.e. the interactions between spatial elements in the landscape: organisms, materials, water); size and shape of patches (i.e. relatively homogeneous areas (same habitat) that differ from their surroundings); connectivity and corridors (i.e. the intensity and spatial dimension of relationships between patches); and buffer zones (i.e. patches that protect sensitive areas from human disturbances and pollution). Just as ecological systems have a spatial component, so too do ecological processes, notably food webs and biogeochemical cycles. The cellular automata approach discussed above has been used to model spatial ecological processes, such as the movement of species or dispersal of materials. Other models have compared different ecosystem management strategies, for instance small dispersed versus large concentrated cuts in forest harvesting.

Landscape ecology takes a grand perspective at a local or sometimes regional scale, using models and GIS. It thus allows for concrete interaction with economic land-use oriented models, in particular for linking the different spatial scales at which models in ecology and economics are commonly defined. The case study considered in later chapters of this book is an example of this. Landscape ecology can provide useful information for ecosystem management, land-use planning and biodiversity conservation. For the last, the size and fragmentation of habitats can be studied. A relevant theoretical perspective is offered here by island biogeography (MacArthur and Wilson, 1967), which predicts the number of species found on an oceanic island, or 'terrestrial island', given its size and distance from the mainland (or other nature areas). Another perspective is based on the notion of metapopulations, spatially separated subpopulations that are mutually connected through the dispersal of organisms. Over time, local subpopulations may become extinct, while also (re)colonisation may occur. Fragmentation of landscapes stimulates the formation of metapopulations, and extreme fragmentation may lead to isolated, non-viable subpopulations (Fahrig and Merriam, 1985; Hanski and Gilpin, 1997). Important conclusions are that nature reserves should be large, circular and well connected with other areas when the aim is for maximum biodiversity. Turner (1998) presents a short introduction to landscape ecology.

Regional and grid-based (GIS) approaches can be linked with landscape and spatial ecosystem ecology. For this purpose, elements can be taken from spatial economics, economic geography, urban economics and social geography. The use of GIS can then support a grid-based linking of, for example, hydrological, ecological and economic processes (as in the case study of this book). This can form the basis for identifying spatial patterns and their evolution. Applications are usually done at an ecosystem or regional scale, rather than at a global scale, because of barriers relating to the level of detail, and the dimensions and size of models. GIS can be combined with analytical models in various ways, ranging from analytical integration to loose coupling based on consistent scenarios (as in the present study). Relatively few integrated models are actually analytically integrated in a GIS format, as they would become very complex and multidimensional. For these reasons, extension towards dynamic formulations is even more limited.

In conclusion, the issues of spatial scale and level are important for integrated modelling. Economics usually distinguishes between micro-, regional (meso-) and macrolevels. Microeconomics describes concrete activities (a household, a business) and their behaviour, even if often through the use of representative agents. Regional and macroeconomics are more specifically oriented towards larger real-world spatial entities: actual provinces, states, regions, nations or the world. Macroeconomics is sometimes also used to denote an alternative approach to microeconomics, namely describing (aggregate) relationships between aggregate variables. To some extent, this implies a trade-off between geographical coverage and descriptive detail. In ecology, a similar distinction can be made between approaches linked to various aggregation levels and spatial scales: global ecology of the biosphere, ecosystem ecology, population ecology at community and population levels, autecology at population and individual levels, and ecophysiology at the individual level. Although economics usually adopts a larger spatial scale than ecology, both are able to address the entire range of spatial scales, from micro to global. Moreover, within a particular economic geographical region (e.g. a country), one may find many homogeneous ecosystem areas, and vice versa. Nevertheless, economics has traditionally been oriented towards larger-scale issues, playing at the national or international level. The rising interest for economic analyses of global environmental change has allowed economics to 'benefit' from its traditional bias. Spatial units often studied in the context of wetlands are a watershed (i.e. a naturally bounded area drained by a river) and a riparian zone (i.e. an area of natural vegetation along a water stream) (see Claessen *et al.*, 1994). Economic approaches still need to be adapted to address such small-scale issues from an economic perspective. The most useful and likely options for integration, from the perspective of limiting model complexity, would probably imply an intermediate level such that, inevitably, some concessions are made in terms of economic and ecosystem model comprehensiveness.

3.4 Integrated modelling and monetary valuation

The previous section showed that integrated modelling of ecosystems has some tradition. The same holds for monetary valuation of ecosystem change (Barbier *et al.*, 1994; Costanza *et al.*, 1997a; Turner *et al.*, 1999; Section 2.3.2). Actual and potential interactions between these two areas, modelling and valuation, can be classified as follows.

- Values estimated in a valuation study can be used as a parameter value in a model study. Value transfer can be used to translate value estimates to other contexts, conditions, locations or temporal settings that do not allow for direct valuation in 'primary studies' (because of technical or financial constraints). Value transfer and meta-analysis were discussed in Section 2.3.2.
- Models could be used to generate values conditional upon particular scenarios; dynamic models can be used to generate a flow of (net) benefits over time, which can subsequently be transformed into a (net) present value. These data can serve as a value relating to a particular scenario of ecosystem change or management.
- Models could be used to generate detailed scenarios that enter valuation or other (stakeholder) experiments. An input scenario can describe environmental change, regional development, a policy or ecosystem management. This can be fed into a model calculation, which in turn provides an output scenario with more detailed spatial or temporal information. These data can then serve, for example, as a hypothetical scenario for valuation, which is presented to respondents in a certain format (graphs, tables, stories, diagrams, pictures) so as to inform them about potential consequences of the respective policy or change.
- Spatially disaggregated models can aggregate monetary values defined at the level of a certain disaggregated spatial patch or unit. This can

support, for instance, a cost–benefit, cost-effectiveness or multicriteria evaluation at the total system level.
- The output of model and valuation studies can be compared. For instance, when studying a scenario for wetland transformation, one can model the consequences in multiple dimensions (physical, ecological and costs–benefits) and aggregate these via a multicriteria evaluation procedure, with weights being set by a decision maker or a representative panel of stakeholders. Alternatively, one can ask respondents to provide value estimates, such as their willingness to pay for not experiencing the change. If such information is available for multiple management scenarios, then rankings based on either approach can be compared.

Integrated models of wetlands have some special features in addition to the general and spatial elements discussed in the previous sections. First, they need to combine hydrological and ecological information, which usually requires an explicit and detailed spatial approach. Second, they usually address a multitude of functions and uses, so that a multiple use approach becomes relevant (Braat, 1992; van der Ploeg, 1990). A representative overview of different approaches is contained in Costanza *et al.* (1997a).

The loss of wetland is often underestimated by decision makers, who base their decision on short-term, visible, directly measurable benefits, such as agricultural output. The loss of wetland functions, such as flood control and barrier to storms, is not fully considered. For example, the Irish Peat Board stated that the ecological damage caused by peat digging is a small price to pay for an improved standard of living (Maltby, 1986, p. 11). The 'value' of a wetland itself and the 'value' of its alternative use for agriculture are often measured as the sum of the direct monetary value of the goods produced on the wetland (e.g. fish) and on the agricultural land (e.g. wheat), without taking into account indirect values. The latter are more difficult to measure in monetary terms and are, therefore, often 'forgotten'.

Several years ago, the US Environmental Protection Agency set up a forum of experts to discuss ecosystem evaluation methods and the information they provide to decision makers. Bingham *et al.* (1995) present an overview of the suggested issues that need to be solved to obtain information that can support proper decisions on wetlands and wetland management. Important information is often missing because the ecological and economic consequences of alternative management strategies are unknown, and therefore also the value of those changes. In order to improve information, first the actual effect of a certain strategy needs to be known, and then its economic value can be assessed. Integrated ecological–economic modelling can support the first part of the problem; it allows insight into the interactions within an ecosystem, especially focusing at the impact of ecosystem changes on functions.

The economic value of a change in a wetland can be compared with, for instance, the costs of abating agricultural pollution in wetlands. Byström (1998) estimates the costs of nitrogen pollution when a wetland is used as a 'natural denitrification plant'. These construction–abatement costs can be compared with other measures to reduce pollution, such as extending fallow periods or reducing nitrogen fertiliser. For the region in Sweden examined, the results show that wetland construction is less costly than reducing fertiliser but more expensive than fallow periods.

The focus of the study described in later chapters is on changes in wetland ecosystems as a result of nature policy and management. Various authors have argued that, as opposed to traditional environmental valuation problems, such as air pollution in cities, valuation of changes in ecosystems requires that account is taken of ecosystem structure, hydrological and ecological processes and multiple services and goods simultaneously provided by ecosystems (e.g. Barbier, 1994; Gren *et al.*, 1994; Turner, 1988; Turner *et al.*, 1998b). The first step in assessing the value of changes in wetlands is to identify general categories of present or future costs and benefits, followed by a selection of specific categories that are most relevant for the cost–benefit analysis in the case study.

3.4.1 *Cost categories*

The following general cost categories can be distinguished: *continuous costs* associated with maintaining one or more wetland functions; *investment costs* associated with maintaining wetland functions; and *opportunity costs* associated with foregoing other uses.

Continuous costs

Continuous costs are associated with maintaining one or more wetland functions and concern regular conservation costs intended to ensure sustainable use of the wetland functions.

Investment costs

Investment costs associated with maintaining one or more wetland functions. include the costs of acquiring areas and the costs of restoring them. They also include the costs of ensuring sustainable use of a wetland function, for example the costs of sanitation of a riverbed and the costs of reducing the phosphate concentration of the water in lakes.

Opportunity costs

In the present context, an opportunity cost of foregone alternative land use and cover is the highest value of alternative land uses. A special type of opportunity costs is addressed by the Krutilla–Fisher adaptation to cost–benefit analysis, which considers lost future values of nature (option and quasi-option value, bequest value and non-use values) as a result of irreversibility of land-use changes. Lost benefits of nature conservation are, therefore, included as opportunity costs of land development (for a clear account, see Porter, 1982).

3.4.2 *Benefit categories*

In the wetland analysis described in the later chapters, irreversibility is not initially emphasised, since we focus on a sustainable nature conservation project. Therefore an ordinary cost–benefit evaluation can be performed and a number of general benefits categories can be distinguished: *actual benefits*, *benefits associated with future use values* and the *costs of replacing a wetland function*.

Actual benefits

The benefits generated by wetland functions can be categorised as use and non-use values; an alternative terminology is active versus passive use values. Use benefits can be further divided into direct use benefits, indirect use benefits and option benefits. Note, however, that there is not yet a general agreement as to the precise meaning of all categories. For example, option value can be considered as both a non-use and a future use value. Estimating benefits of uses is not as straightforward as estimating costs of maintenance of functions. Following Barbier (1994), direct use benefits have been derived from direct use or interaction with a wetland's resources and services, and they can be valued by market analysis based on the travel costs method, and non-market valuation using contingent valuation. Consumption uses can often be valued via market-based methods. Indirect use benefits reflect the indirect support and protection provided by wetland functions and can, among other approaches, be valued by estimating the costs of replacing these functions with human-made technology (Gren *et al.*, 1994). Non-use benefits can only be valued by non-market valuation methods.

Benefits associated with future use values

The optimal method for assessing future values remains unresolved. First, consider option and quasi-option values. Option value reflects the uncertainty of individuals about their future demand for goods and services depending on wetland functions, while quasi-option value reflects the value of preserving options for future use given some expectation of the change in information or knowledge. Second, consider (anthropocentric) non-use benefits. These reflect individuals' knowledge and appreciation of the wetland, provided they do not have any intention of using the wetland in any way. If this is motivated by altruistic considerations for other species, the term 'existence value' is used. In addition, 'philanthropic value' denotes considerations for other human beings that are currently alive. Finally, if the appreciation applies to future generations, then the term 'bequest value' is used.

Costs of replacing a wetland function

Assessment of the costs of replacing a wetland function can serve as an indicator of the lower bound on the economic value of the associated wetland function. These replacement costs can be included as *benefits* in the cost–benefit analysis. An example of estimating the economic value of a wetland by the costs of replacing its functions with human technology can be found in Gren *et al.* (1994). One of the problems of this approach is that such replacement only represents a partial substitute for the wetland, since it cannot replace general ecosystem functions such as climate regulation and the maintenance of the stock of biological resources. The fundamental reason for this problem is that where economic production processes usually serve a single function, ecosystem processes are usually multifunctional.

Note that there is a potential problem of double-counting. When separate economic activities use or are based on the same function(s) generated by a wetland, valuation of each independently can result in double counting of benefits. How can we deal with this problem? An option would be to omit certain estimated benefits in order to make sure that the benefits associated with a particular wetland function are counted only once. One has to be careful, though, as by omitting certain estimated benefits the remaining sum of benefits may underestimate the total value associated with a particular wetland function.

Table 3.2 lists information about possible wetland costs, benefits and methods in order to assess these for a hypothetical example that shows much similarity to the one presented in the case study of this book. Note that the term project is appropriate here as it indicates that a change is valued. Indeed, the change can be interpreted as an investment project that aims to contribute to a situation of sustainable nature conservation. The changes concern land/water use and land cover and imply an enlargement of the surface area of the wetland ecosystems in the area. The investment costs encompass acquisition and restoration of areas, and the sanitation of rivers and lakes. The change in land/water use and land cover also induces changes in the costs of maintaining wetland functions and

Table 3.2 *Categorisation of possible activities, benefits and costs for a hypothetical wetland change*

Human activity/ changes	Benefits associated with the wetland function(s)	Costs associated with the wetland function(s)	General estimation method (uncertainty indication)
Nature conservation	Non-use value and use value Sub-categories: bequest value, existence value and (quasi-) option value	Continuous costs and investment costs associated with maintaining these functions	Benefits estimated via stated preference by means of non-market valuation methods (uncertain)
Nature-based recreation	Use value Sub-categories: direct use value, indirect use value and option value	Continuous costs and investment costs associated with maintaining these functions	Sum of net recreational benefits (estimated by market analysis), and added value derived from consumer surplus (by non-market valuation) (uncertain)
Agriculture	Use value Sub-category: direct use value based on market analysis; non-market techniques for subsistence/ artisinal production	Continuous costs and investment costs associated with maintaining these functions	Added value estimated via market analysis of agriculture in the area (reliable)
Function replacement: replacing the wetland function with human-made technology to provide protection against storm and flooding	Use value Sub-categories: indirect use value and quasi-option value	Costs of replacing a wetland function with human-made technology	Costs of replacement technology (uncertain)

Table 3.2 (cont.)

Human activity/ changes	Benefits associated with the wetland function(s)	Costs associated with the wetland function(s)	General estimation method (uncertainty indication)
Function replacement: replacing wetland functions with human-made technology to provide a nutrient sink	Use value Sub-categories: indirect use value and quasi-option value	Costs of replacing function with human-made technology Maintenance costs to compensate for unsustainable use of functions (sanitation of the bottom of the river, reducing phosphate concentration in lakes)	Costs of replacement technology serve to indicate added value of the associated wetland function (uncertain)

changes in costs to ensure sustainable use of a wetland function. In addition, it induces changes in benefits derived from wetland functions.

3.5 Performance indicators

3.5.1 *The role and use of indicators*

An indicator is a measure describing a system or process in such a way that it has a significance beyond its face value (Lorenz, 1999; Lorenz *et al.*, 2001a,b). Indicators translate data into information pertinent to their users. Environmental indicators tend to be used by authorities responsible for environmental management. Indicators are also capable of structuring the diversity of information needs and flows among different elements of society (e.g. different management authorities, the public and politicians).

Indicators can be used in a variety of ways. Their most common use is descriptive. Since indicators tend to be measured regularly, the resulting time series can be used to show trends in system performance. However, this depends on consistency in data collection and indicator construction. An indicator can also be compared with a reference condition that represents a desired state. Environmental standards (e.g. Adriaanse, 1993), sustainability criteria (e.g. Azar *et al.*, 1996; Gilbert and Feenstra, 1994) and conditions prior to disturbance (e.g. Ten Brink *et al.*, 1991) provide examples of this use of indicators.

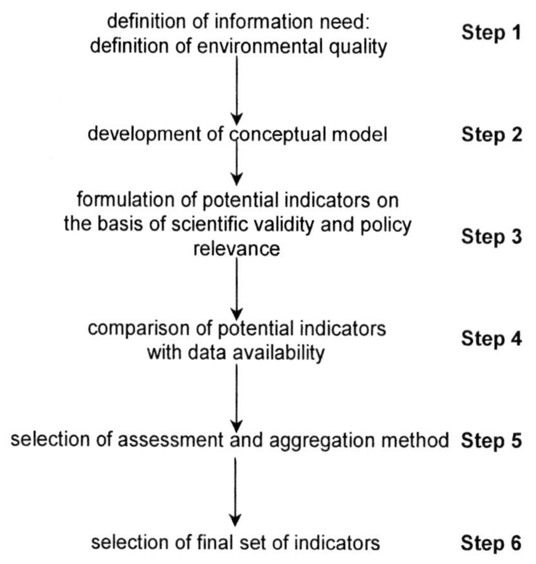

Fig. 3.1 Guideline for the development of indicators. (Adapted from Verhallen, 1995.)

In addition, indicators can provide information on possible future developments. Simulation modelling facilitates experimentation with virtual futures. The linking of such models with indicators offers a number of advantages (Rotmans, 1997):

- illustration of links among indicators;
- insight into the dynamic behaviour of indicators;
- long-term projections;
- identification of critical variables and guidance for indicator selection;
- calibration of models and presentation of output; and
- development and evaluation of policies and related costs.

3.5.2 *Indicator development*

Indicators are often presented without specification of the selection process. Even when ranking and weighting schemes for selection are presented, it is often not clear how judgments have been made in their development. A guideline for indicator development and selection is shown in Fig. 3.1 (Lorenz, 1999; Lorenz *et al.*, 2001a; Verhallen, 1995). The six steps from this guideline are briefly discussed below.

Step 1. Definition of information need

Specification of information need depends on the particular phase in the policy cycle. In general, an environmental problem goes through a policy life cycle with different phases: problem specification, policy development, policy implementation, policy control and policy evaluation (Winsemius, 1986). The policy phase for the management and ecological rehabilitation of wetlands differs among countries and even among specific wetlands, and so the information need also differs. The case study presented in this book can be placed in the policy development phase.

Step 2. Development of a conceptual model

A conceptual model is a verbal or visual abstraction of a part of a world from a certain point of view. It contains information on the system, such as its spatial and temporal scale and interactions among components, so that the particular problem to be solved is represented (Bakkes *et al.*, 1994). Specification of the concept of environmental quality (Section 2.2.6) provides an example of a conceptual model.

Step 3. Formulation and evaluation of potential indicators

The dominant variables in models (whether conceptual or quantified) provide the basis for potential indicators. While the literature offers a large number of criteria on which indicators could be evaluated (Lorenz, 1999), priority is placed here on their scientific basis and policy requirements (Kuik and Verbruggen, 1991; Liverman *et al.*, 1988; OECD, 1994; Swart and Bakkes, 1995; van Harten *et al.*, 1995). Additional criteria depend on the aim of the indicator and the policy phase of the issue in question. Problem definition indicators need to be sensitive to environmental change, show rapid response and indicate long-term trends. Indicators within the policy development phases could focus on the success of existing policy, or on possible gains from alternative policies. Within the policy implementation, control and evaluation phases, indicators would permit comparison of present and target states (Heij and Bannink, 1995).

Step 4. Data availability

Data are available in a variety of databases, such as those managed in the Netherlands by the Central Bureau of Statistics (CBS, 1996a). Additional data may be obtained from specific sources (e.g. economic data in the case study described in this book was provided by the Institute of Agricultural Economics (LEI)), the literature, monitoring programmes, surveys and special projects. Conversion of these data to the appropriate spatial scale can pose problems. They are usually collected for administrative regions and subsequently aggregated (for example, to the national level), whereas they are often needed at scales corresponding to natural processes or geographical features, such as river basins.

The dependence of indicator development on data can lead to the situation that data availability drives the selection of indicators, which, in turn, reinforces the collection of the same data. Water-quality monitoring systems have been an example of the 'data-rich but information-poor' syndrome, in which many data are produced but not tailored to information needs (Ward *et al.*, 1986). Guidelines for the design and implementation of monitoring systems (Adriaanse *et al.*, 1994; Cofino, 1994) bear a strong resemblance to this guideline for indicator development.

Step 5. Selection of assessment and aggregation method

Value functions can be used to assess indicator values. Value functions provide an explicit link between factual information (e.g. the indicator value) and a value judgement ('good' or 'bad') in the form of a mathematical representation (Beinat, 1995). A value function translates the performances of management alternatives into a value score, which represents the degree to which a management objective is met. The value is a dimensionless score, usually ranging between 1 (best available performance) and 0 (worst performance).

It may not be possible to specify indicators that reflect an evaluation objective (such as a performance indicator for environmental quality). The direct effect may be unknown or the right data unavailable. Proxy indicators may be developed; for example, the area of riparian vegetation could approximate the quantitative relationship between river rehabilitation measures and species diversity. The use of proxy indicators mixes the use of factual information (e.g. the area of riparian vegetation) and a value judgement (e.g. the support that riparian vegetation provides to river species). However, these indicators should still describe the objective as closely as possible (Beinat, 1995).

Reference conditions are useful to relate the indicator to a value judgement. They act as anchors for interpreting the meaning of a value score. Consider the use of chlorophyll concentrations as an indicator for nutrient pollution in a lake or river. This indicator could have the value of 1 at natural background concentrations, a value of 0 at concentrations when algal blooms are present and 0.5 for an intermediate situation. The form of the value function depends on the change in value judgement with the increasing (or decreasing) indicator value. Common forms are block functions, linear, sigmoid, convex or concave curves.

Reference levels can be absolute or relative, leading to an absolute or relative judgement, respectively. For example, functional standards enable an absolute judgement on the suitability of a resource for a certain piece of goods or supply. The 'no observed effect concentrations' are thresholds for absolute ecological effects on species in an ecosystem. Relative references enable only a relative valuation: relative in time or space, or relative to policy aims. The selection of an absolute or relative reference depends on the aim of the assessment and on the availability of data on reference levels. Relative references are often used when no absolute references are available.

Aggregation of a number of variables into one value implies the steps of selection, weighting, scaling (transforming indicators into dimensionless measures) and mathematical manipulation. Aggregation methods such as weighted averages and weighted summation require that there are no interdependencies and no double-counting among indicators (Beinat, 1995). For indicators of environmental quality, the strong interdependence among ecosystem processes makes it difficult to ensure that neither of these problems occurs.

Step 6. Selection of final set of indicators

Data availability and selection of methods for assessment and aggregation lead to the final set of indicators.

The next question is how to go from indicators to performance indicators? Linking models and indicators permits the comparison of alternative futures and thereby the assessment of the strategies that could lead to these futures. Gilbert and Feenstra (1994) show the power of such a link when indicators and models are combined with scenario analysis and specification of a desired future state. Their sustainability indicator could be used to assess the success or failure of alternative policies with regard to cadmium use and the subsequent accumulation of cadmium in the soil. The linking of indicators to models, and using this system to assess alternative futures, creates a particular category of indicators, termed performance indicators. Performance indicators translate model output into scores for criteria on which alternative scenarios or management strategies may be evaluated. As such, their role is to support decision making.

The specific characteristics of performance indicators are that they provide information on the effects of management strategies, they are based on model output and they respond as directly as possible to an evaluation objective. The selection of performance indicators from the set of potential indicators is driven by:

- theory: such as the specification of environmental quality as an evaluation objective;
- the case study and the issue being investigated;
- model capability and output: essentially the issue of data availability;
- the existence of suitable reference levels for indicator assessment; and
- the avoidance of double-counting and interdependencies when aggregating.

In the case study presented later in this book, performance indicators are generated for three objectives: economic efficiency (the conventional evaluation objective), environmental quality and spatial equity.

3.6 Evaluation

An evaluation method can support the ranking of alternative decisions regarding management, policy, development scenarios or projects. The method can provide a complete ranking, the best alternative, a set of acceptable alternatives or an incomplete ranking. Most integrated studies deal with multiple performance indicators with specific units. Then one can use either monetary valuation of indicators in physical units (changes) in combination with cost–benefit evaluation, or multicriteria evaluation of indicators in monetary, physical and other units. Since in the present study no use will be made of monetary valuation techniques or results, this section will focus on multicriteria evaluation (for a comparison, see Hanley, 1999, Janssen and Munda, 1999).

In general, multiple, different criteria are relevant in the evaluation procedure. Each of these should be comprehensive and measurable. Comprehensive means that the criterion's value is sufficiently indicative of the degree to which a specific objective is met. Measurable means that it is consistent with a particular measurement scale that allows ordering of the alternatives (management strategies, policies, scenarios, projects) for a particular objective. The ratio, interval ordinal and binary scales qualify as suitable scales of measurement, as they make it possible to order alternatives.

There is a wide range of available multicriteria evaluation methods for continuous and discrete alternatives, based on graphical methods, using different aggregation procedures, based on ranking of the importance of criteria (lexicographic methods), etc. (Janssen, 1992; Nijkamp *et al.*, 1990). Janssen and Munda (1999) present a short overview of multicriteria evaluation in the context of environmental issues. In practical applications, two types of input are required for a multicriteria analysis: (i) a table with effects scores for each alternative and criterion (or indicator) and (ii) a set of weights. Sensitivity analysis can focus on either of these. Multicriteria evaluation can be regarded as a more general method of welfare analysis, and this makes it a good alternative to traditional, neoclassical economic analysis of social welfare and policy instruments (see Munda, 1997).

Many integrated modelling and assessment studies are based on explicit dynamic and spatial models that generate dynamic or spatial patterns for indicator variables. These are generally difficult to compare without aggregation, over indicators, space and time. Only cost–benefit analysis includes an explicit, standard procedure to solve this problem via discounting. This has been much debated in economics and environmental economics (e.g. Price, 1993). Multicriteria analysis can include the time or spatial pattern of effects by specifying evaluation criteria that reflect the different spans of the effects. Via this route, short- and long-term effects can be included in a single effects table and decision rule. Not all multi-criteria methods are suitable for dealing with spatial and temporal issues (see Nijkamp *et al.*, 1990). Furthermore, uncertainty, which is often crucial and made explicit in integrated modelling studies, can receive a special treatment in multi-criteria analysis (see Munda, 1995). More generally, when dealing with dynamic indicators for evaluation criteria, one can aggregate these in various ways. One approach is to discount each indicator pattern according to a specific procedure; even when only monetised values are included, varying discount rates can be used for separate variables each measured in monetary terms (compare with the Krutilla–Fisher algorithm; Porter, 1982). Alternatively, one may regard each data point (i.e. for a given indicator and point in time or space) as a separate attribute in the multicriteria evaluation procedure and assign weights to each of these. Again, this is also possible when dealing only with monetised values. This approach will not be very attractive in general, as it may require the choice of a large set of weights, especially when temporal and spatial dimensions are considered simultaneously, which is clearly impractical. Still other approaches can focus on average or final indicator values over time or space.

3.7 Conclusions

In this chapter, a broad overview has been given of integrated assessment and modelling. In this context, integration can refer to integration within the field of natural sciences (notably earth and biological sciences), as well as between natural and social sciences. Integration is characterised by many different approaches. Differences can be attributed to *ad hoc* factors (coincidences), modelling techniques adopted and contributing disciplines. They also relate to specific choices made with regard to model characteristics, such as generality, precision, realism and disaggregation. The observed variation is consistent with the large number of conceptual approaches to integration of natural and social science identified in Section 2.4. Spatial modelling has received special attention as it is of importance in integrated modelling in general, and in the case study described in this book in particular. Spatial modelling can benefit from the inclusion of various elements of spatial economics and landscape ecology, as well as from methods like GIS and cellular automata.

The use and generation of monetary indicators in integrated modelling was also discussed. A number of potential interactions were identified between modelling and valuation, including support of parameter settings, value projections under particular scenarios, generation of value scenarios for experiments, aggregation of values with models, and comparison of modelling and valuation studies. So far, these areas have been rather independent or ever antagonistic, but the discussion here suggests that benefits of synergy are possible. Finally, performance indicators and evaluation were discussed. These topics are often omitted in similar studies to the one described in this book and the reader is left wondering how to deal with an overflow of information generated by complex, detailed model exercises. Here concrete suggestions are made for the design and selection of indicators in a single framework. Various aspects of multicriteria evaluation were briefly discussed, as this method seems most suitable to address multidimensional output of integrated models.

4

Theoretical framework and method of integrated study

4.1 Objective and approach

The Vecht area study is intended to contribute to the development of an integrated framework and method for multidisciplinary analysis of policy and management scenarios for areas dominated by wetland ecosystems. The nature of wetlands, characterised by transitions from land to water, is such that several disciplines need to provide information and insights to allow evaluation of the impacts of proposed changes in wetlands areas. These changes include, for example, land reallocation, management of water tables, regulation of economic activities and construction of infrastructure. As discussed in Chapter 2, the core disciplines in wetlands research are hydrology, ecology and economics. Hydrology studies qualitative and quantitative aspects of stocks and flows of groundwater and surface water. Ecology studies the relationship of living organisms with their environment. Economics studies the behaviour of households and companies in the context of market interactions between the demand and supply of goods and services. As discussed in Chapter 3, these disciplines can be integrated for the purpose of case studies through the use of integrated economic–environmental modelling. The approach adopted in the Vecht area study is consistent with the very basic and general drivers–pressured–state–impact–response (DPSIR) framework that was discussed in Section 2.4. Chapter 5 will present each of the elements of this framework in the context of the case study.

The study involves a system of static, spatial models that are externally linked to arrive at a complete regional model of relevant hydrological, ecological (vegetation) and economic characteristics. In particular, three types of model are interlinked (Fig. 4.1). A spatial hydrological model of groundwater and surface water is based on a mix of statistical regularities. It describes both the

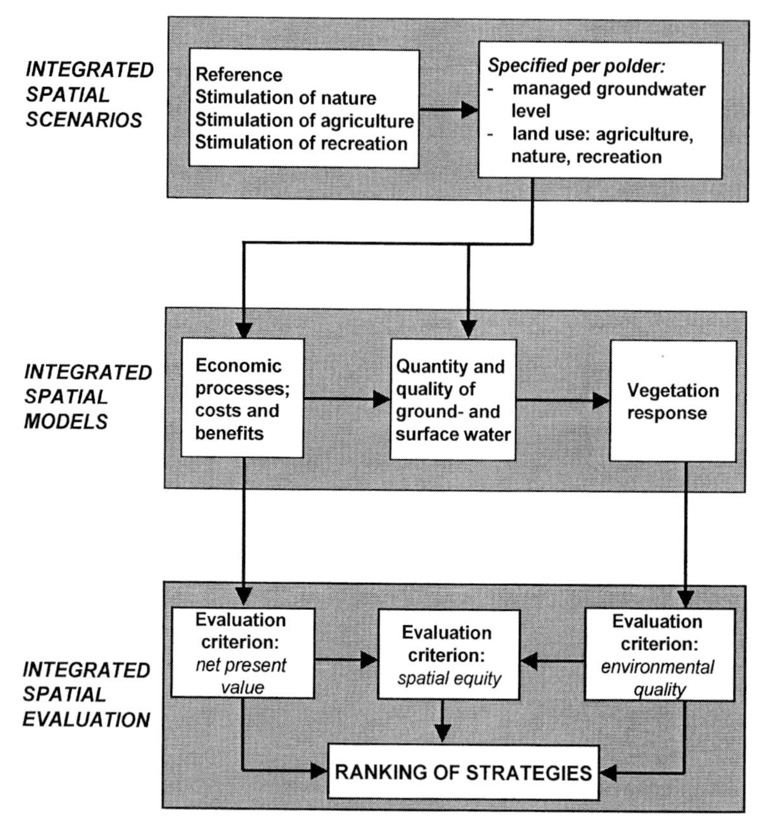

Fig. 4.1 The integrated approach followed in the Vecht area case study.

quantity and quality of water. The abiotic indicators coming out of this model enter a vegetation response model. This model has a statistical predictive format. It transforms a set of abiotic factors into probabilities of the occurrence of plant species. These can be considered as representing medium-term equilibria, given certain management conditions in wetlands (e.g. mowing). An economic model describes a number of core activities (agriculture, nature conservation, outdoor recreation) in terms of costs and benefits as well as other economic indicators (such as employment) associated with particular activities and regulatory policies. In addition, some environmental indicators are generated, which depend on the magnitude and intensity of agriculture. A more detailed explanation of the model structure is presented in the next section.

Given the static integrated modelling approach, the scenarios are static as well. They will be used in a comparative static analysis, where changes relative to a reference (or base) scenario are compared among alternative scenarios. This may be interpreted as reflecting stable conditions and is consistent with the static character of the integrated model approach. From an economic perspective, however, the approach reflects a finite time horizon, since reinvestment of activities is needed after a period of between 10 and 30 years. The same holds for (parts of the) semi-natural environmental system, the characteristics and processes of which cannot be maintained without human intervention but need restoration at regular times.

4.2 Three integration levels

The method of integrated study addresses integration at three levels, as shown in Fig. 4.1. The first level involves the formulation of development scenarios for the Vecht area that include consistent settings for the hydrological (e.g. water levels) and economic (e.g. intensity of agriculture) parameters, reflecting combinations of land use and water management that are feasible in reality. The second level entails the integration of hydrological quality and quantity modelling, vegetation response modelling, and economic modelling and accounting. The hydrological–ecological part of the modelling and analysis is performed on a grid basis and subsequently aggregated to the level of polders, to make it consistent with the economic analysis. Integration at this level includes explicit links between the natural science and the economic models. This focuses on the inclusion of nutrient surpluses from agriculture in the water quality modelling, and on the influence of water quantity and quality on plant species occurrence. A third integration level consists of two steps. The first one aggregates output from the models into performance indicators reflecting objectives in the evaluation. The final and essential step in the integration is the evaluation procedure in which these performance indicators are used to rank scenarios. Three objectives drive this evaluation: net present value (or economic efficiency), spatial equity and environmental quality. The last attempts to capture ecological criteria describing how well wetland ecosystems are functioning (see Section 2.2.6).

The scheme in Fig. 4.1 summarises the interactions (reflected by arrows) between the various elements of the integrated study framework.

1. Scenarios are defined on the spatial grid and polder level. They include land use and cover, settings of water levels, type of economic activities and some additional policy settings. The scenarios are extreme in the sense that they reflect clear choices with regard to physical planning, nature policy, agricultural policy and recreation incentives. Three specific scenarios are defined, aimed at stimulating or giving priority to nature, agriculture and recreation, respectively. The model results are subsequently calculated and presented as changes with regard to a

reference or base scenario, which is defined as a static extrapolation of the present situation (sometimes referred to as 'business as usual').

2. The economic model generates a set of indicators for a number of activities on a polder level. These indicators include net present values of costs and benefits, nutrient outflows and employment. A scenario defining land use, water level and economic activity type and intensity on a polder level determines the main input to the economic module. The approach can be adapted easily to address feedback from ecological indicators to additional nature recreation benefits resulting from improved environmental quality (biodiversity). Some illustration of this is given in Chapter 7. Nutrient outflow indicators serve as input to the hydrological model. The cost and benefit indicators enter the evaluation procedure.
3. The hydrological model for groundwater and surface water stocks, flows and quality is steered by information from the economic module (nutrient flows) and scenarios (initial groundwater levels). A regional groundwater flow model of the research area has been constructed based on a general groundwater model. Given initial conditions, the model can predict quantities of groundwater flow throughout the region. From the model results, the groundwater chemistry or quality is calculated (i.e. abiotic factors are predicted).
4. The hydrological model generates output that enters the ecological model ICHORS ('influence of chemical and hydrological factors on the response of species'). This is a vegetation response model that has been statistically estimated on the basis of a large set of empirical observations in the area. Given conditions of abiotic variables at a specific site, such as morphology of the surface water system, depth of the water, hydrology and water chemistry, ICHORS estimates the probability of 265 types of wetland plant species at that site. Evaluation of the model output allows the identification of the presence of groups of species that are characteristic for specific types of wetland vegetation. The vegetation response probabilities enter the evaluation procedure.
5. The evaluation procedure is aimed at ranking the scenarios. As input, it uses the economic net present value and environmental quality indicators. These can be defined and calculated at various spatial scales, depending on the particular type of evaluation purpose (cost-effectiveness, equity between subregions, equity between subpopulations, etc.). Economic indicators include employment and net present values of changes in activities. Environmental quality is based on three general indicators: process, structure and resilience (Section 2.2.6). Multicriteria analysis is

used to perform an aggregate evaluation of these indicators. This includes checking for dominant scenarios and pairwise comparisons of trade-offs between indicators. Spatial equity will be studied for gains and losses (combining economic and environmental indicators) on the level of polders.[1]

In Fig. 4.1, the two arrows from 'integrated spatial scenarios' to 'integrated spatial model' reflect the consistency between economic and hydrological parameters arranged via the scenario formulation. The arrow going from the first to the second box in 'integrated spatial model' reflects the influence of agriculture on abiotic conditions in the groundwater. The second arrow here denotes the influence of abiotic conditions on vegetation probabilities. The three model types in combination with the degree of spatial disaggregation imply that a dynamic system is both too complex and too data demanding. Therefore, the integrated system of models has a static character. In terms of analysis, the approach can best be characterised as a comparative scenario analysis.

4.3 Spatial dimension

The analysis has an explicit spatial dimension. The hydrological model is formulated on a grid base. The economic model is defined at the level of polders, so it generates a spatial distribution of economic indicators on a polder level. The various natural science and economic variables at grid and polder scales will be used to construct three sets of indicators: economic, environmental quality and spatial equity. These indicators are aggregated via different procedures to arrive at an overall ranking of scenarios.

The scenarios are spatially disaggregated and are formulated at the level of grids and polders. This contributes to both the accuracy and the realism of the descriptions. The hydrological parameters are defined at a grid level (500 m × 500 m) and the economic parameters at a polder level (on average 200 ha). This is related to the fact that the hydrological model assumes homogeneity at the level of grids, and the economic model at the level of polders. Any information at a grid level can be easily aggregated to the polder level. Each scenario specifies land use and water levels at the polder level for the 73 polders that make up the study area.

From a policy perspective, the detailed spatial character is useful as it allows the analysis of spatially refined scenarios. In particular, the nature and

[1] Earlier studies of the Vechtstreek have focused on cost–benefit analysis of specific scenarios and policy measures (Barendregt *et al.*, 1992; Bos and van den Bergh, 2002).

recreation scenarios attempt to improve environmental quality in a north–south corridor through the study area.

4.4 Relevance for policy and management

The operational policy approach takes the form of development scenarios that can be implemented into the models. The choice of scenarios reflects extreme policy choices constrained only by physical and biological realities, as well as by very strict constraints in present planning and policy.

The results can inform policy at three levels: national, regional and local. Spatial planning and water management are of interest to various public authorities at all three levels. In addition, the interaction of physical planning (land-use policies) and hydrological management can be examined. Until now, such policies were often separately analysed as well as separately designed.

The study presents a tool for testing whether the national government can preserve the international value of nature in the Vecht area and simultaneously maintain its economic characteristics and values. In particular, it can examine whether the Ramsar status of the wetland reserve 'Naardermeer', located in the northern part of the study area, implies that a larger buffer zone is needed. This is also relevant from the perspective of the ecological network ('Ecologische Hoofdstructuur') of the Netherlands and the EU Habitat Directive in the European ecological network NATURA 2000. The most important question relates to the future of agriculture in the area. In this context, coordination of the policies of the two provinces that govern the area is relevant, as they have different opinions about agricultural development: the spatial separation of activities and nature versus the interweaving of nature and agriculture. This also has to be evaluated against the historical context and social desirability of the cultural landscape. Finally, various local institutions can profit from the findings and the method presented here. The most important ones are the water boards and the nature preservation and management organisations, notably 'Natuurmonumenten' (a non-governmental organisation) and 'Staatsbosbeheer' (the forestry service part of the Ministry of Agriculture, Nature and Fisheries).

5

The Vecht area: history, problems and policy

5.1 Description of the area

The Vecht area is a plain with wetlands, located between the river Vecht in the west (Fig. 5.1) and the sandy ice-pushed hill ridge 'Het Gooi' approximately 8 km to the east. To the north, a former sea ('Zuiderzee'), now reclaimed, bounds the area; 20 km to the south near the city of Utrecht, the hill ridge and the river almost meet (Fig. 5.2). The area is a wetland region with many shallow lakes. Even where a solid soil is present, the groundwater table is close to the surface. The soil diversity is represented by the presence of three types of soil: close to the river and the former sea, there are deposits of clay soils; the centre of the plain is covered by a peat layer some metres in depth; and in the eastern direction this peat layer gets thinner until it reaches the sandy soil of the hill ridge. The area straddles two provinces (North-Holland and Utrecht) and includes 10 municipalities. The number of inhabitants in these areas is restricted. The centres with significant population sizes are mostly at the border of, or just outside, the area (see Table 5.1).

Agriculture and nature dominate land use; industry is almost absent (Fig. 5.3). Agriculture is primarily dairy farming. About half of the surface is covered with pastures. Most of these are wet and located on the peat soil with groundwater about 40 cm below surface. Surface water management in the polders is necessary to maintain these pastures. Ditches at a precision level of centimetres manage the water tables. The pastures are important not only for agriculture but also for nature: many wading birds nest here, characteristic plant species exist and various aquatic species are present in ditches and canals.

Most nature areas are owned by nature conservation organisations. The variety in natural and artificial lakes, reedbeds, marshes, grasslands and alder forests

Fig. 5.1 The river Vecht with nature at the shoreline and recreation on the water.

creates a mixture of different succession phases characteristic for these types of wetland (Figs. 5.4 and 5.5). The value of nature in this area is high, both from a national and an international perspective. One lake area, the Naardermeer, is listed as a Ramsar wetland and also belongs to the Dutch Ecological Network. In addition, outdoor recreation generates values of nature and landscape. For instance, recreation includes sailing boats on the river Vecht and some of the lakes, housing, camping sites and substantial use of the roads and paths for walking and cycling. In particular, many people from the cities of Utrecht, Amsterdam and Hilversum regularly visit the area.

5.2 Historical development of the area

Thousands of years ago, the study area was uninhabited. The wetland landscape, in its pristine condition, was very different from the present landscape. Various landscapes have existed at various points in time. The river Vecht was originally a free branch of the river Rhine. It already had its present bed and a flood zone where clay was deposited and on which nutrient rich forests could grow. In the centre of the plain, there was an enormous area of bog vegetation, fed by rainwater and with a higher water table than the river. At the border of the hill ridge, groundwater – infiltrated in the ridge as rainwater – seeped up and

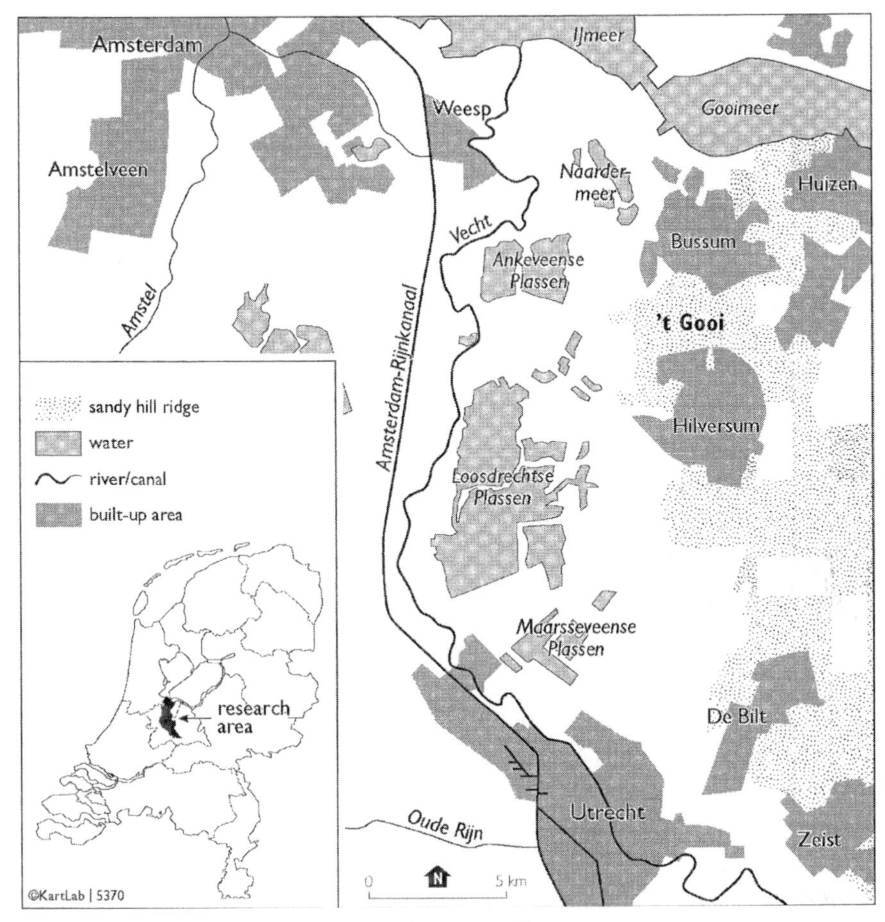

Fig. 5.2 The research area and its surroundings.

created constantly wet, buffered and mesotrophic conditions. Sedge vegetation dominated the landscape at such locations.

When the first people arrived in the area, they reclaimed the peatlands by making small ditches; this allowed the upper soil to be drained. In the fourteenth century, many areas were used for growing crops. In the sixteenth and seventeenth century, the increasing populations of Amsterdam and Utrecht required fuel. This was provided, in part, by digging peat from the bogs (containing oligotrophic *Sphagnum* peat). This led to an enormous change in the landscape. Initially, large peatlands were dug to the phreatic level; later the peat was dredged below the surface water table to the sandy subsoil level. Peat was dug up in areas of about 50 m × 200 m with the help of a long-handled tool. To dry the (wet) peat for use as fuel, it was deposited on the remaining baulks of uncut peat (*legakker* in Dutch). The wet peat was trampled to make it more solid, and it was then

Table 5.1 *Inhabitants and surface area of municipalities in or near the Vecht area (January 1996)*

Municipality	Number of inhabitants	Area (km^2)
Located (at least partly) in the Vecht area		
Breukelen	13 912	43.7
's-Graveland	9 390	21.7
Hilversum[a]	83 272	46.0
Loosdrecht	8 971	15.1
Maarssen[a]	41 172	28.3
Maartensdijk	9 392	39.7
Muiden	6 804	14.3
Naarden[a]	16 637	21.5
Nederhorst den Berg	5 216	11.6
Weesp[a]	18 073	20.7
Located within 15 km of the Vecht area		
Almere	112 704	132.1
Amersfoort	114 884	56.0
Amstelveen	75 869	40.6
Amsterdam	718 119	160.4
Bussum	30 893	8.1
De Bilt	32 562	26.9
Diemen	23 464	12.0
Huizen	41 351	15.9
Nieuwegein	59 214	29.7
Soest	42 801	45.9
Utrecht	234 254	61.4
Zeist	59 188	48.5

[a] The majority of the inhabitants live outside the Vecht area, as described in the text.
Source: CBS (1997a).

cut and dried over several months to make it suitable for transport to towns. Numerous turf ponds (*petgat* in Dutch), separated by 5–10 m high baulks, have remained in the landscape (Fig. 5.6). The uncut strips of peatland were small since it was profitable to dig as much peat as possible. Sometimes the strips were so small that wave activity in the turf ponds completely washed them away. As a result of this erosion, which became more powerful once various turf ponds had merged, many large lakes developed during the eighteenth and nineteenth centuries. One of these lakes was reclaimed in the nineteenth century, resulting in the present polder Bethune, an agricultural area with pasturelands. In the same period, a natural lake in the centre of the Vecht area was reclaimed,

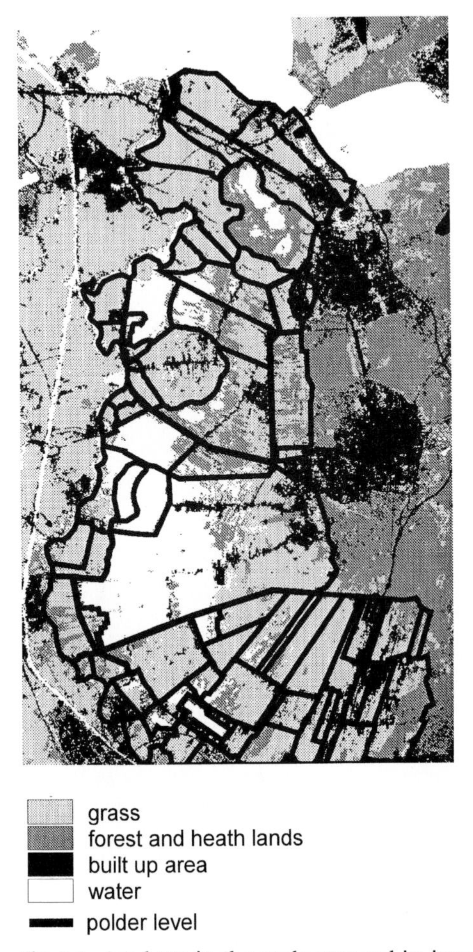

Fig. 5.3 Land use in the study area and its immediate surroundings.

leading to the formation of the Horstermeer polder, which is also used for agriculture nowadays. However, the natural lake Naardermeer, in the northern part of the Vecht plain, has been preserved, in spite of two efforts to reclaim it.

In the seventeenth and eighteenth centuries, the landscape at the borders of the Vecht area was reshaped as well. People from Amsterdam, who had become rich from trade and shipping, established estates close to the river Vecht. This gave rise to a large number of estates with extensive gardens in the agricultural landscape. The western border of the hill ridge Het Gooi underwent a similar change. Here the sandy soil was dug away in the seventeenth century in order to construct the (nowadays famous) canals in the centre of Amsterdam. The profits of this venture allowed a large number of estates to be established in the

Fig. 5.4 Succession in a fen area from open water via aquatic vegetation and reedbeds to alder forests.

area from which the sand originated. Today, these estates are famous for their landscape and architectural beauty.

Until the middle of the twentieth century, the 'functions' agriculture, nature and housing in the Vecht area were quite balanced. The intensity of agriculture and human activities was rather low, and bird and plant species had a high diversity. However, in the first decades after the Second World War, drastic changes took place. Recreation was developed in the area, leading to an increase in sailing boats, notably on the lakes by the village of Loosdrecht. Industrial activities arose just outside the Vecht area, resulting in the pollution of surface water within the area. Several locations in the area with turf ponds were used to dump industrial waste. Drinking-water companies abstracted so much water in

Fig. 5.5 Mesotrophic quacking fens with a species-rich vegetation of sedges and herbs.

Fig. 5.6 The characteristic landscape of the Vecht area with turf ponds covered with water lillies.

Fig. 5.7 Agriculture in the Vecht area: pasturelands surrounded by nature.

the hill ridge that the original supply of groundwater to the wetlands dried up. Moreover, to maintain the water tables in the summer, polluted water from the river Vecht was let into the area. In this period, agricultural management also intensified, resulting in more cattle and more manure. The manure was spread on the pastures, which connect to nature areas through water channels. This increased the outflow of nutrients into the nature areas (Fig. 5.7).

5.3 Threats to the wetlands

The problems related to the wetlands can be summarised as relating to hydrology, chemistry and physical planning. The first problem concerns the fact that surface water and groundwater influence the hydrology of the wetlands. Groundwater and rainwater are the original sources of water in the wetlands. The average amount of rainfall has not changed over recent decades, but the input of groundwater has decreased drastically. During the 1970s, 17×10^6 m^3 of drinking water were abstracted from the hill ridge on a yearly basis, resulting in lower water tables in the hill ridge and a reduction of the seepage flow of groundwater into the wetlands. Because of high evapo-transpiration in the summer and the risk of mineralisation in peatlands with low water levels, surface water had to be supplied from the river Vecht to maintain the water tables. There are similar hydrological problems near the two deep polders (Bethune and

Horstermeer), which drain their surrounding areas through groundwater flows into these polders. This has caused the original groundwater flow from the hill ridge to the river (in an east–west direction) to be reversed into a flow of surface water from the river to the wetlands (in a west–east direction).

The second problem, that of water chemistry, is related to the previous processes. The wetlands have become polluted through the supply of polluted water from the river Vecht into the wetlands. The concentrations of phosphorus and nitrogen in the surface water have increased and, consequently, algal blooms have occurred. Not only has the river been a source of nutrients, but also increased agricultural management, local pollution from sewage water treatments, mineralisation of peat soils and outflow from illegal waste-dumps have contributed to nutrient flows. As a result, the quality of nature in the area has deteriorated. The algal blooms have resulted in the disappearance of certain types of aquatic vegetation. The natural process of ecosystem succession from an aquatic vegetation to an early terrestrial phase with floating mats – the habitat of a number of important 'red list' (threatened and protected) species – has come to a halt (Fig. 5.8). Because of the poor buffering capacity of groundwater in the root zone of the vegetation and the increased acid deposition through air pollution, acidification has also become an important problem in recent decades.

The third problem stems from the pressure caused by human activities, though these are regulated by physical planning. Recreation in the centre of the area has intensified, especially during the 1960s. New boatyards and camping sites were created, in spite of the official status of the area being mainly 'nature'. These changes have occurred everywhere in the area, although the local municipalities have attempted to regulate the various activities via physical planning. Farmers have tried to intensify their activities but have not been very successful because of technical restrictions on water tables and physical planning regulations to protect nature. Many farms have now even disappeared. Recently, a very restricted number of farms have been allowed to locate in the areas between the nature areas. The remaining farmhouses have often been used for other purposes, especially for housing by rich people from outside the area.

As a result of all these changes, the natural characteristics of landscape have significantly altered. The originally undisturbed area with extensive agricultural management has been replaced by a landscape interrupted by various types of human activity. Moreover, the area available to pure nature has decreased in size. Although most nature areas are still present, connections between them are disturbed by the various human activities and accompanying physical infrastructure. As a result, many nature areas have become too small to support viable plant and animal populations in the long term.

Fig. 5.8 The cormorant, one of the characteristic species in the region, present in a colony of around 10 000 pairs.

5.4 Economic activities

During the twentieth century, agriculture was the main economic activity in the Vecht area. Although in terms of land use this is still the case, the economic profits of agriculture arise mainly in the northwestern and southern parts of the area. In the central part of the plain, where nature predominates, the number of full-time working farmers is small.

Recreation developed during the 1930s and expanded during the 1950s and 1960s. Nowadays, the river Vecht is one of the major waterways for recreation boats in the western part of the Netherlands. It connects the river Rhine and its secondary rivers to the lake IJsselmeer. Some lakes used for recreation are connected via sluices with the river Vecht. The lakes at Loosdrecht are especially famous for sailing; on a day in the peak season, thousands of boats can be found on these lakes. Moreover, the presence of many boatyards, restaurants, hotels and camping sites has created a very popular location for recreation and tourism. The entire Vecht area is well known for its pleasant cycling and walking routes. Many people from Amsterdam, Hilversum and Utrecht, as well as many smaller towns, spend their time here enjoying the landscape and nature. Many secondary roads are used for cycling, and various smaller paths for walking. The pubs and restaurants in the villages in the area particularly benefit from this type of recreation. At a small number of locations, it is possible to participate in excursions into nature areas, arranged by nature conservation organisations.

The presence of a fine landscape with many lakes at such a short distance from the large cities means that the Vecht area is an attractive place to live, especially when a boat can be moored close to home. Since at present physical planning does not allow the construction of additional houses, existing houses in the area, notably those along and close to the lakes, have become extremely expensive. On average, house prices are about 30% above those in the rest of the Netherlands.

Trade and industry are severely restricted by physical planning. Inside the wetland area, only a few small factories are present. Most other industries are located along the river Vecht and along a canal at the eastern border of the area.

5.5 Policy and management in the Vecht area

Government at various levels impinges on the state and changes of nature in the Vecht area. Public policy and management are embedded in a great number of decisions and policy plans, relating directly to regulations under the heading of 'nature policy' or indirectly to regulations under the headings of 'environmental policy', 'physical planning' and 'agricultural policy'. Policy can be oriented towards specific activities, such as recreation, nature or water use, or towards the combination of different activities in specific areas via land use and physical planning. Policy is structured at three levels: national, regional and local. At the national level, decisions are very general but overruling; policy at the lowest (local) level is the most detailed. Management in the area is consistent with plans at the various levels and is carried out by municipalities and provinces, together with various other stakeholders, such as nature organisations and economic interest groups.

5.5.1 *International and national policies*

The national government incorporates the international value of nature in the Vecht area via the Ramsar status of the wetland reserve Naardermeer. This reserve is also one of the assigned areas in the EU Habitat Directive. The entire Vecht area is in the list of 'Areas Important for Birds' (EU Birds Directive), which is supposed to be integrated with the Habitat Directive in the European ecological network NATURA 2000.

In the physical planning of the national government (VROM, 1990), the Vecht area has been assigned the 'green course' label, a classification for national development purposes implying that the area should remain rural or natural. Here, ecological qualities define the direction of development: natural ecosystems should be preserved and stimulated, and natural hydrological patterns and the quality of the environment should be improved. In the national nature policy plan (LNV, 1990), the entire Vecht area is defined as part of the main ecological network of the Netherlands. This approach requires that nature reserves in the area are protected and that farms in areas between nature reserves contribute to stimulating the development of nature in agricultural areas. This is also reflected in the National Water Management Plan (V&W, 1989). The river Vecht and the surrounding areas greatly influence the hydrological processes in the area. The National Water Management Plan addresses the problem of lowering water tables, especially in peat areas, where mineralisation of the organic soil is regarded as a serious risk. The policy plan of the non-urban areas in the Netherlands adds two other aspects (LNV, 1993). First, it stresses the importance of rural and natural areas for recreation. Plans have been presented to add several woodlands to the densely populated western parts of the country. Second, it notes that the wet conditions in the Netherlands should be used to stimulate water-dependent ecological relations in the Netherlands. This has been done by defining a 'Blue Axis' in a national project, Netherlands–Waterland.

Physical planning by the two provincial governments (North-Holland and Utrecht) corresponds with the national policy but is far more detailed. It defines areas amenable for nature and agriculture, and those suitable for the construction of new buildings and for economic development. Although there is a tendency to preserve the landscape, because of both its cultural–historical and nature value, the provinces also try to meet the vested interests of the inhabitants, trade, industry, recreation and infrastructure. The two provinces have different opinions about agricultural development in the Vecht area. The province of Utrecht strives as much as possible for the physical or spatial separation of economic interests from nature areas. As a result, there are relatively few restrictions on agricultural management in the Utrecht part of the Vecht area. In contrast, the province of North-Holland aims to establish an area where nature and agriculture are interwoven, with the basic premise that nature and agriculture are part of the cultural landscape. As a result, agriculture is faced with relatively many restrictions.

At the local level, 10 municipalities have a policy corresponding with the provincial policy. Nevertheless, local decisions by the municipalities, for instance to build new houses for their inhabitants, may conflict with higher level decisions. The community of Maartensdijk, which covers a large part of the province of Utrecht in the Vecht area, has a policy to reduce urban development and to stress the importance of nature and agriculture. In Loosdrecht, economic pressure from recreation has led to less-stringent policies. In former periods, many nature areas were transformed into boatyards, marinas and camping sites, and the owners of these represent the vested interests in these locations. North of Loosdrecht, in the province of North-Holland, the interweaving of different functions is important, especially in the central parts with many nature reserves. In the northwestern part of the area, agriculture is more competitive. This is reflected by the policies of those communities, which are intended to take care of farmers and the conditions they face.

5.5.2 *Regional and local activities*

However, without further policy and management, the Vecht area cannot be considered as following a sustainable development path. There is a conflict between the policy goals discussed above and the existing local activities and management. The total number of stakeholders in the area is very high. Nature preservation and management organisations, such as Natuurmonumenten and Staatsbosbeheer, have very detailed plans (for each parcel). Agricultural areas are managed by the local farmers, restricted by national regulations. Surface water tables and surface water chemistry are managed by water boards, which have their own regulations and taxes. The environmental problems in the Vecht can only be solved with the help of area-based management that takes account of the great number of conflicting interests, activities and plans at different levels and considers the different stakeholders. Below, three specific regional management activities are discussed in more detail.

Abstraction of drinking water

In the hill ridge Het Gooi, east of the Vecht area, in 1975 more than 17×10^6 m^3 of groundwater was abstracted for drinking-water purposes. This groundwater creates very favourable hydrological conditions in the wetlands as it is fresh, has the correct chemical composition, is unpolluted and seeps out of the hill ridge into the wetlands at a constant rate. However, using it for drinking water causes a shortage of water in the wetlands. This shortage has been compensated by the inlet of polluted water from the river Vecht, but this has resulted in the eutrophication of the wetlands. After a long period of debate and various studies, the province of North-Holland finally decided that groundwater abstraction should be limited to at most 7×10^6 m^3 per year. This meant that the abstraction wells had to be reallocated, and that other sources of drinking water had to be sought.

In the western parts of the lakes at Loosdrecht, there is a storage basin of drinking water for the city of Amsterdam. Formerly, this basin obtained its water from these lakes, but since the lake water became polluted another source has had to be used; this is surface water from the Bethune polder, located to the south of these lakes. This polder supplies Amsterdam with 25×10^6 m^3 of water per year. It is quite deep, 4 m below sea level, which causes water to drain in from the surrounding polders and the hill ridge Het Gooi. The resulting water shortages in those polders had to be compensated in order to maintain the water tables needed for nature (and even agriculture). Surface water from the river Vecht was used for this purpose, but this gave rise to water pollution problems in the polders. A sustainable solution might be to inundate the Bethune polder so that a structural change in hydrological patterns is achieved. However, the responsible government – in this case the province of Utrecht – decided that the production of drinking water for the city of Amsterdam was more important than the stimulation of nature surrounding this polder.

Restoration projects

Much energy and money has been expended in the Vecht area since the 1960s on environmental issues. The main problem is the eutrophication of the surface water, resulting in algal blooms and other unpleasant conditions for recreation. Various restoration projects have been initiated over time to reduce these problems, mostly by the responsible authority: depending on the issue, a province, a municipality or a water board. Examples of projects during the last 15 years are:

- reallocation of the outlets of sewage water treatment plants;
- dredging of aquatic sediments with high nutrient loads from lakes;
- application of sewage water pipes all over the area;
- introduction of new sources of water supply (e.g. phosphate removal from water);
- recycling of unpolluted water from polders without water shortage to polders with water shortage; and
- increasing surface water tables to stimulate nature development.

Despite these projects, many problems remain. The resilience of ecosystems has been difficult to predict, partly because it is characterised by hysteresis. Nevertheless, the interest in solving environmental problems has increased over recent years. As a concrete success, restoration projects have put a halt to the reduction in quality and quantity of nature areas during recent decades.

Nature development

In addition to the restoration of nature, a number of projects aim to develop nature in areas previously dominated by agriculture. These areas are bought by the government or nature conservation organisations and then reshaped to result, whenever possible, in natural conditions. This is done to enlarge the nature areas in accordance with the national policy goals. As a result of such management strategies, the Ramsar wetland Naardermeer will be surrounded by new nature reserves. These will be formed of flooding agricultural lands that were reclaimed at the beginning of the twentieth century.

In the southern half of the Vecht area, there are a number of important reserves with many 'red list' (threatened and protected) species. The maintenance of these reserves is difficult as they are small and surrounded by agricultural areas. To create favourable conditions, active management in surrounding previous pasturelands, hydrological stimulation and physical planning are combined. These scattered reserves have recently become part of one large nature reserve.

Another way to stimulate nature is through financial support by the government to farmers at selected locations to help them to attune their management to support the wading birds that breed in the pastures.

5.5.3 *Land reallotment*

The national government has facilities to reconstruct rural areas. The instrument of land reallotment has been used during recent decades to stimulate conditions in agricultural management. This has improved the water tables, led to the creation of new roads, reestablished farm houses at strategic locations and facilitated the exchange of land parcels in order to provide farmers with parcels close to each other and, whenever possible, connected. During the last few years, this instrument has also been applied for other purposes, including nature stimulation and the creation of recreational infrastructure. Land reallotment takes a long period (10–15 years) because of the necessity for detailed planning and consultation with each landowner in the area. The final plan has to be approved by the owners of the land. The advantage of such an approach is that all authorities and stakeholders in the area are involved in changing land use in parts of the area. The approach also feeds back to the physical planning, which province and municipalities can change to support land reallotment. The procedure results in a plan in which both economic and environmental management are updated on the basis of the most recent information and technology. Financial support by the government is provided to implement the plan.

The part of the Vecht area located in the province of North-Holland is presently undergoing a land reallotment procedure. The preparation for this started in 1994. The part of the Vecht area in the province of Utrecht, with the exception of the polder that includes the lakes of Loosdrecht, had undergone such a land reallotment procedure in the early 1990s. The province of Utrecht attempted to stimulate the rural economy in this area by separating the conflicting uses of agriculture and nature in single polders. Some nature areas were enlarged with areas where formerly agriculture had dominated, while water tables were allowed to rise. In other parts of the area, the conditions for pastureland were improved so that agriculture could become more intensive and thus yield higher profits. The lowering of water tables is one aspect of this latter policy.

5.6 Conclusions

The Vecht area is a wetland area with many functions, sometimes at a single location. Nature is an important one; in some reserves it is the only function, in other parts of the area it is one among multiple functions. The same applies to recreation, with intensive recreation being concentrated in certain parts of the area. Extensive recreation, such as cycling, is possible throughout almost the entire area. The cultural value of the landscape and the appreciation by visitors is high. Historically, the area was predominately in agricultural use, giving rise to a unique landscape that was maintained for centuries. However, over the last few decades, the economic activities in the area have changed. Some direct benefits from nature to farmers, such as fishery, mowed reed used for the roofs and peat digging, have disappeared. Many farmers have gone out of business and the new owners of their farms use the pastureland merely to keep horses or sheep. Only in the northern and southern parts of the area is agriculture profitable, partly because of support by subsidies from the European Union. The profitability at these locations mainly accrues through the less-strict regulations, which allow for high water tables. The lowering of water tables will give rise to peat mineralisation and possibly the subsidence of the foundations of houses in the area. The inhabitants of the area are aware of these risks and generally support maintenance of the wet conditions through high water tables. In other words, the conditions for sustaining the wetlands are very much in line with the economic viability of the area.

Public policy and private activities in the Vecht area are directed at providing sufficiently favourable economic conditions for the inhabitants and at the same time maintaining the presence of animal and plant species in the wetland landscape. From an international perspective, the values associated with the wetlands are significant, at least according to the Ramsar Convention and the EU Habitat Directive. This is recognised in national and provincial physical planning. Moreover, nowadays, the construction of new buildings is very restricted, new settlements of industry are prohibited and the preservation of the cultural heritage in landscape and estates predominates in physical planning. Finally, in both provinces associated with the study area, the development of facilities for recreation is being stimulated.

The problems of pollution and low groundwater levels need to be solved in order to obtain sufficient conditions for a sustainable wetland environment. During the 1990s, many private activities as well as the abstraction of groundwater have been brought in line with this goal. The conflicts between agriculture and nature have been solved in various ways. In some locations, the conditions for nature have been improved, while in others the development of new nature areas is being directly stimulated. In addition, subsidies are given to farmers in order to restrict agricultural management and promote nature.

The Vecht area is located in a densely populated region of the Netherlands. It appears possible that many economic activities can occur in the Vecht area in the presence of a number of restrictive institutional arrangements that assure minimum conflicts with the wetland functions. The current approach is that the vested interests of the present stakeholders, such as the inhabitants of villages, the owners of sailing boats, farmers, nature conservation groups and drinking water companies, should be integrated as much as much as possible with wetland preservation. The stakeholders, mostly well organised, already participate in the relevant policy discussions. Only when all stakeholders agree on a set of solutions can the resilience of the wetland area be achieved on a permanent basis. The next chapter will formulate development scenarios for the Vecht area based on its characteristics as described in this chapter.

6

Development scenarios for the Vecht area

6.1 Introduction

This chapter presents development scenarios for the Vecht area that will be used in the integrated modelling exercise reported in later chapters. These scenarios reflect choices made in physical planning, nature policy, agricultural policy and the regulation of recreation and drinking water production, which are the main political and economic interests in the area. The scenarios have a spatial disaggregation and are formulated at polder level. The hydrological parameters are defined at grid level (500 m $\times$ 500 m) and the economic parameters at polder level. This is because the hydrological model assumes homogeneity and is formulated at grid level, while the economic model is formulated at the polder level. Any information at grid level can be aggregated up to polder level.

As the static model approach is used, static scenarios are formulated. Each scenario will provide information about land use and water levels at polder level for 72 polders (Fig. 6.1). The procedure for such a spatially disaggregated approach is complicated. Land use depends on the type of activity and the type of soil. The first includes historical conditions, classified into agricultural land (grassland, cropland), nature and forestry, urban settlements and open water. The soil type is classified into sand, clay and peat. The scenarios also define certain basic economic conditions. In the context of the Agriculture scenario, this includes the choice between intensive and extensive agriculture, which, in turn, will influence the amount of nutrient flows, drainage and productivity. In the Nature scenario, this includes conversion of agriculture to nature. In the Recreation scenario, investments in and revenues of land- and water-based recreation are considered. The consequences of the hydrological changes (from the

Fig. 6.1 The study area and its polders. Each polder is identified by a number; the corresponding names are given in Table 6.2 (p. 115).

Reference or Base scenario) are that water flows alter, resulting in changes in the relative proportions of surface and groundwater. Given the different chemical characteristics of these water types, this will lead to water quality changes in different parts of the area. The physical characteristics of the soil and existing infrastructure provide constraints, as do spatial interactions and spatial buffering zones.

Table 6.1 *Development scenarios for the Vecht area*

Scenario	Hydrological settings per polder	Economic settings per polder
I. Reference	Present situation in all polders	Present situation in all polders
II. Stimulation of agriculture (three types of polder setting)	0 = no change in water table 1 = no change in water table 2 = –0.2 m	0 = no change in land use 1 = 50% conversion of present to intensive agriculture 2 = 100% conversion of present to intensive agriculture
III. Stimulation of nature (three types of polder setting)	0 = no change in water table 1 = +0.1 m 2 = +0.2 m	0 = no change in land use 1 = 50% of present agriculture converted to nature 2 = 100% of present agriculture converted to nature
IV. Stimulation of recreation (five types of polder setting)	0 = no change in water table 1 = +0.1 m 2 = +0.2 m 3 = polder flooded 4 = no change in water table	0 = no change in land use 1 = 50% of present agriculture converted to nature, and investments for outdoor recreation 2 = 100% of present agriculture converted to nature, and investments for outdoor recreation 3 = polder flooded and opened for recreation 4 = water-based recreation stimulated or intensified on existing open water

The four scenarios for the development of the Vecht area are:

I. Reference;
II. stimulation of agriculture (Agriculture);
III. stimulation of nature (Nature); and
IV. stimulation of recreation (Recreation).

The details of these four scenarios are geven in Table 6.1. The scenarios focus on particular policy changes, while taking into account the present conditions. Scenarios II to IV focus on core human activities in the region and allow comparison of quite distinct, although still realistic, future organisations of the Vecht area. Each of the scenarios is defined and explained below. Table 6.1 summarises the scenario settings. Figures 6.2–6.4, below, show the spatial composition of scenarios II, III and IV, respectively. Note that the information in the figures reflects changes relative to the Reference scenario (I).

6.2 Scenario I: reference

In this Reference (also known as the Base or Business-as-usual) scenario, land use in all polders is assumed not to change in the future. This scenario serves as a benchmark to compare results obtained under the other scenarios, and in particular to derive results under each scenario in terms of changes relative to the reference situation. The Reference scenario shows the different types of land use and their geographical mix in the area (see Fig. 5.2 in Ch. 5). A central corridor through the Vecht area comprises both nature reserves and agricultural activities (pastureland). Nature reserves include lakes, reedbeds, alder forests and various types of species-rich grassland. In the western and eastern parts, agriculture dominates. Along the river Vecht itself and at the extreme eastern border with the sandy hill ridge, there are many estates from the seventeenth century. Villages are scattered throughout the area. Recreation activities take place at various locations in the region. Water-based recreation is particularly important for the river Vecht and the lakes at Loosdrecht.

6.3 Scenario II: stimulation of agriculture (Agriculture scenario)

Under the Agriculture scenario, water management is directed toward fulfilling the needs of agriculture as it is spatially distributed at present (Fig. 6.2). The composition of agriculture changes but the total area with agriculture remains the same as under the Reference scenario. The aim is to change the structure of agriculture in a way that leaves profits intact, or even increases them. This means that the amount of nutrient run-off can increase, thus leading to additional environmental pressure. The present nature reserves are maintained without extra buffer zones, and so the boundary between the intensified agricultural areas and nature reserves will be very sharp. No restrictions on agricultural activity would apply (e.g. on fertilizer use). A distinction is made between the different settings for (average) agricultural intensity, which is consistent with the hydrological parameters. This means that a higher (lower) density of cows

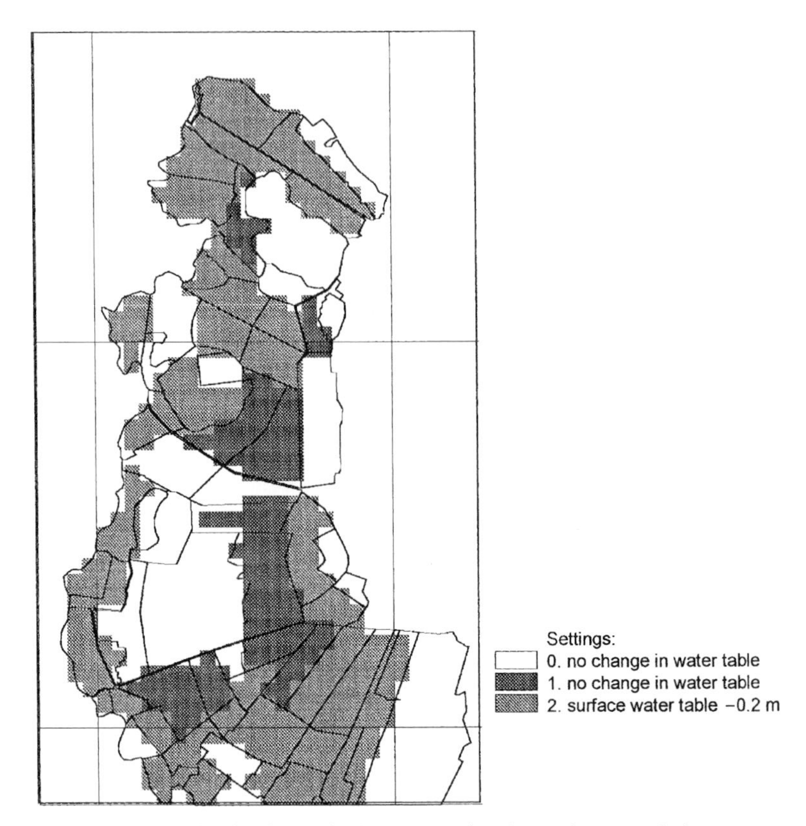

Fig. 6.2 Stimulation in the Agriculture scenario: shows changes relative to Reference scenario.

per hectare of cultivated land and lower (higher) water tables reflects a higher (lower) intensity of agriculture. Details about the different types of agriculture are provided in Chapter 8. More specifically, the settings for the 73 polders are as follows.

0. No changes occur relative to the reference situation.
1. The water table is not changed; 50% of agricultural land is intensified.
2. The transition to areas with 100% agriculture to optimise the agricultural economy. One element is the lowering of the managed polder water level by 0.2 m relative to the reference situation. No restrictions are put on the intensity of agriculture.

6.4 Scenario III: stimulation of nature (Nature scenario)

Nature conservation and restoration is stimulated under the third scenario (Fig. 6.3). At least 50% of previously agricultural land is transformed into areas available for nature conservation, in order to guarantee the viability of ecological systems and functions. Agriculture, recreation and housing cannot expand and have to satisfy certain physical constraints. These include restrictions on land drainage, on fertilizer use and on dealing with wading birds. In some polders, agricultural areas will be transformed into nature management areas: agricultural activities will be stopped or relocated. The remaining pasturelands will change into nature areas. Finally, this scenario includes a non-polder-specific element: dephosphorisation of water from the river Vecht so that this source of water for the region will have nutrient levels within acceptable boundaries. Finally, three dephosphorisation plants are part of this scenario to allow for a direct beneficial effect, allowing for economic costs, on water quality.

More specifically, the 73 polders will have particular settings in this scenario:

0. No changes occur relative to the Reference scenario.
1. The managed water tables go up 0.1 m to support the restoration of nature. Half of the agricultural land in a polder is converted to nature. Subsequent land management (implying land use) is of two types. First,

Fig. 6.3 Stimulation in the Nature scenario: shows changes relative to Reference scenario.

some polders are managed for mixed nature and agriculture management with priority given to nature. In these polders, the integration of nature and pastureland is stimulated to maintain the landscape and the characteristic wetland ecosystem. Second, agriculture and recreation are integrated with landscape and nature. Polders with this form of land management tend to constitute a mosaic of agriculture and estates from the seventeenth century; this generates much landscape value.

2. To stimulate the wetlands, the water level is raised by 0.2 m. All agricultural land in a polder is converted to nature. This includes certain specific situations: nature reserves already in existance will become 100% nature and land currently under agricultural use will return to natural wetland. Because of the change to high water tables, agriculture will no longer be profitable.

6.5 Scenario IV: stimulation of recreation (Recreation scenario)

This Recreation scenario is an extension of scenario III (stimulation of nature) with stimulation of outdoor recreation (Fig. 6.4). Shifts in agricultural practice will also occur here through the adoption of less-intensive – more-extensive – methods. The practical counterpart of this is a voluntary arrangement called Relatienota, under which the government gives a subsidy to individual farmers to maintain the landscape such that it is profitable for recreation. In addition, nature-based outdoor recreation occurs in larger areas, supported by investments to improve accessibility and infrastructure, and economic services oriented towards recreation. This covers both land- and water-based recreation activities. In order to improve conditions for nature as well as for water recreation, some polders are flooded under this scenario. The potential settings for polders are as follows.

0. There is no change relative to the reference situation. This includes (a) land-based activities, where stimulation of nature or recreation is not needed, not wanted, or not possible; and (b) the existing lakes with water recreation.
1. The managed water tables rise by 0.1 m to support the restoration of nature. Half of agricultural land in a polder is converted to nature. These mixed land-use polders are made attractive for outdoor recreation by investing in cycle paths, horse tracks, hotel capacity, nature education centres, etc. Either recreation is allowed in present areas with open water or existing water-based recreation activities are allowed to intensify.
2. To stimulate the wetlands, the water level is raised by 0.2 m. All agricultural land in a polder is converted to nature. These polders are also

made attractive for outdoor recreation by investing in infrastructure and services.

3. Some polders are flooded to stimulate water-based recreation and, indirectly, nature. This involves restoration, investments in services and infrastructure, and change of ownership. A disputable issue is that agricultural activity is surrounded by nature areas. This covers partial flooding of the Horstermeer polder (except for the village), and partial flooding of the Bethune polder to create a recreation area and harbour.
4. Either recreation is stimulated in present areas with open water or existing water-based recreation activities are allowed to intensify. Nothing changes from a hydrological perspective (i.e. water tables remain as in the reference situation).

Fig. 6.4 Stimulation in the Recreation scenario: shows changes relative to the Nature scenario. Note that settings 1 and 2 of this scenario are the same as those shown in Fig. 6.3.

6.6 Conclusions

Although in a spatially disaggregated system, as in this case study, many different scenarios are conceivable, only three – in addition to a reference situation – have been proposed here. These scenarios provide an extreme (though realistic) but possible range of developments for the Vecht area. 'Realistic' should here be interpreted in terms of the various physical (hydrological), ecological, economic and institutional limits at the polder, provincial and national levels. At the same time, the set of scenarios explores the important use and value dimensions of the area: agriculture, nature conservation and recreation. Taken together, these features suggest that analysis of the chosen scenarios can contribute to an understanding of the system and its potential developments.

The scenarios will be used to perform model calculations of changes relative to the Reference scenario. More details on hydrology and nature-related motivations underlying specific hydrological settings in the various scenarios are presented in Chapter 7. More details on economic settings relating to agriculture and recreation are provided in Chapter 8.

Appendix 6.1 List of polders

Table 6.2 lists the names of the polders and the numbers allocated to each. The location of the polders is shown Fig. 6.1.

Table 6.2 *The 72 polders forming the study area*

No.	Polder name	No.	Polder name
1	Noordpolder beoosten Muiden	37	Polder Mijnden oost
2	Binnendijksche overscheensche Berger en Meentpolder	38	Polder Breukelen Proosdij noord
3	Zuidpolder beoosten Muiden	39	Oostelijke binnenpolder van Tienhoven
4	Muiden buitendijks	40	Polder Westbroek west
5	Nieuwe Keverdijkse polder noord	41	Loosdrechtse Plassen
6	Naardermeer	42	Kerkeindse polder
7	Keverdijkse overscheense polder	43	Polder Achtienhoven
8	Nieuwe KeverdiJkse polder zuid	44	Achterweteringse polder
9	Heintjesrak en Broeker polder noord	45	Achterweteringse polder noord
10	Heintjesrak en Broeker polder zuid	46	De Ster
11	Hilversumse Bovenmeent	47	Egelshoek
12	Spiegelpolder	48	Loosdrecht
13	Hollandsch Ankeveensche polder	49	Kievitsbuurt polder
14	Hilversumse Ondermeent	50	Tienhovense Plassen
15	Horn en Kuijer polder	51	Vliegveld Hilversum
16	Hilversumse Meent	52	Polder Bethune
17	Stichtse Ankeveensche Polder	53	Polder Maarsseveen
18	Nieuwe Keverdijkse polder noord	54	Gansenhoef
19	Hornpolder	55	Polder Breukelen Proosdij zuid
20	Horstermeer polder	56	Kievitsbuurt Plassen
21	's Gravelandse polder	57	Molenpolder noord
22	Overmeer	58	Molenpolder
23	Kortenhoefse Plassen	59	Achterweteringse polder
24	Hollandsch Ankeveensche polder	60	Wilgenplas
25	Meer-uiterdijkse polder	61	Maarssen
26	Polder Dorssewaard	62	Kleine Plas
27	Stichtse Ankeveensche polder	63	Ruigenhoekse polder
28	Kortenhoefse Plassen	64	Polder de Kool
29	Wijde Blik	65	Maarseveense Plasssen
30	Het Hol	66	Oostwaard
31	Loenderveen	67	Polder Buitenweg
32	Vuntus	68	Polder Binnenweg
33	Waterleiding Plas	69	Polder Gagel
34	Loenderveense Plas	70	Utrecht
35	Loosdrechtse Plassen	71	Bethune polder
36	Polder Mijnden west	72	Horstermeer polder west

7

The spatial–ecological model: hydrology and ecology

7.1 Introduction

This chapter describes the hydrological and ecological modelling that underlies the integrated modelling approach in this study. In addition, some general characteristics of the intermediate results are discussed. Aggregation of the latter will be pursued in later chapters. Modelling the interaction of hydrology and vegetation of a wetland has three main elements.

1. A water quantity model, describing amounts of groundwater and surface water expected at specified locations in the present hydrological conditions;
2. A water quality model, describing the chemistry of groundwater and surface water at specified locations; and
3. An ecological model describing the presence of wetland plant species at specified locations, depending on local environmental conditions.

These three models interact, as summarised in Table 7.1. The water quantity model provided inputs to the water quality model, while both these models predict the abiotic conditions that serve as input for the ecological model.

A set of boundary conditions is needed to define the domains of the models. These boundary conditions can then be manipulated in scenarios. However, some input variables of the model are assumed to be static in the time domain adopted here, e.g. soil texture.

The modelling has a spatial dimension. Therefore, each of the models requires spatially differentiated input. For the storage and representation of spatial data, a raster or grid representation is chosen. The raster representation is easier (more compact) than a vector representation and allows for more efficient calculations

Table 7.1 *Input and output of the three applied models*

Model	Input	Output
1. Quantitative hydrology	Surface water tables; groundwater tables; boundary conditions in soil	Level of groundwater per cell; water balance per polder
2. Qualitative hydrology	Chemistry of local surface water, groundwater and rain; the water balance from (1)	The chemistry of surface water and groundwater
3. Ecological	Water tables from (1); chemistry from (2); local conditions per polder; some management options	The probability of encountering 265 plant species per cell

of results. A raster representation is also suitable to depict spatially varying continuous variables like surface level and groundwater tables. In a raster representation, a minimum grid cell size in the raster is chosen and assigned one value for a specific variable. Quick and fast overlay computations can be done with these raster maps. For this study, a raster cell size of $500\,\mathrm{m} \times 500\,\mathrm{m}$ is chosen. Each cell is assumed to be homogeneous.

The most important spatial unit in this study area is the polder. Polders are in all cases larger than a raster cell. All spatial data for the modelling are available in raster maps. Some data are digitised from existing paper maps; others are reworked from existing (polygon) data files, and still others are interpolated from point measurements with spatial interpolation techniques. All data are stored in ArcInfo and PCRaster geographical information systems (GIS).

7.2 The hydrological quantity model

The original hydrology in the river plain has been influenced by three sources: precipitation, groundwater and river water (Fig. 7.1). Precipitation is uniformly distributed over the area. Because of the phreatic water table in the valley, the rainwater flows into the river at surface level. In addition, rainwater infiltrates in the sandy hill ridge to become groundwater. This groundwater flows to the lowest point in the area, the valley of the river Vecht. The river can flood the area adjacent to the valley. Nowadays this process is influenced by human activities. A major part of the infiltrated water in the hill ridge is abstracted for drinking water and is not available in the valley. Deep polders in the valley attract groundwater and drain the surrounding area. Since the flow of surface water is fully regulated, resulting deficits in water supply are compensated by

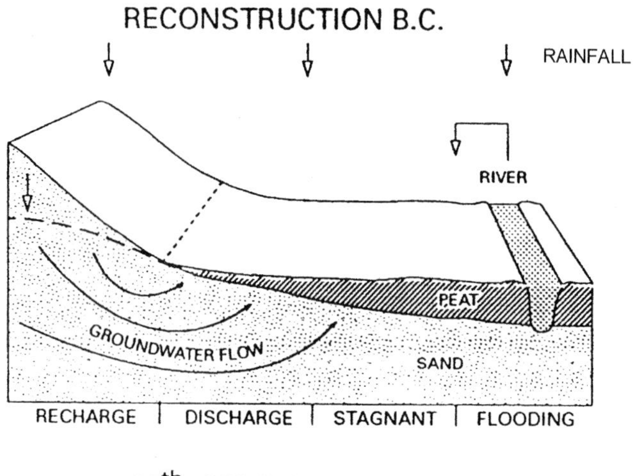

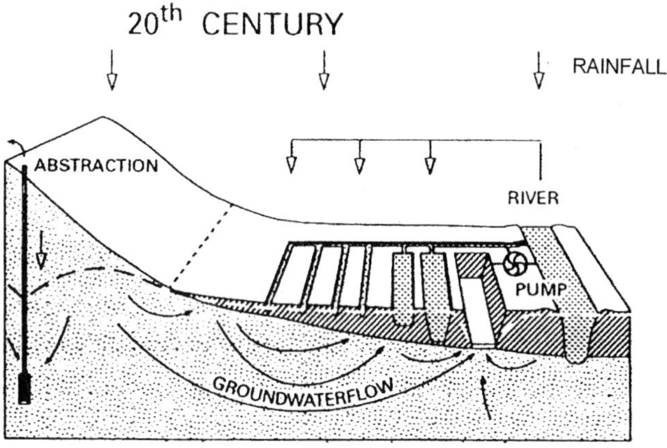

Fig. 7.1 Hydrology of the case study area.

the inlet of river water into the area via pumping. For details, see Witmer (1989) and Schot (1989, 1991).

The flow of groundwater is modelled with the model code MODFLOW (McDonald and Harbaugh, 1984). This code solves three-dimensional flow problems with a finite difference matrix-solving module. This model code has been widely applied (Anderson and Woesner, 1992). Modelling is based on the Darcy flow equation:

$$Q = cA(h_2 - h_1)/L\,,$$

with Q the flux of water, c the conductivity of the soil layer, $h_2 - h_1$ the difference in water level between two points in the model space, A the surface through which flow occurs and L the distance between measure points h_1 and h_2.

The model area is divided into several layers of blocks with the same lateral conductivity. Each block is assumed to be homogeneous in conductivity. In each block, a central point is defined for which the flow is calculated. For a three-dimensional set of model nodes, the model code generates, iteratively, a solution for the differences between each node; that is, a finite difference approach is followed.

For each model, node flows are calculated in the x direction (Q_x), the y direction (Q_y), and the z direction (Q_z). From these partial flow terms, a net flow (Q_{net}) is calculated. Since the model calculates a stationary situation, these flow terms are associated with equilibrium conditions, based on the assumption that there is no change in the storage of groundwater in the soil.

Apart from the flow term between nodes, other sources or sinks of water can be modelled with the code. These are then added to or subtracted from Q_{net}. Recharge (usually as net precipitation) is a net in-term, since the area has a net precipitation surplus. The recharge flux is added to the top layer of the model. Rivers and small ditches can either drain or infiltrate water from the top layer of the model. Small ditches are modelled as sinks ($-Q_{\text{drn}}$), while rivers can drain or infiltrate water in the top layer (Q_{riv}). Abstraction of groundwater for drinking-water purposes is a net out-term ($-Q_{\text{sub}}$) and is put in the model layer from which the water is abstracted.

For each model node, the set of flux terms is added to obtain the final flux (Q^*):

$$Q^* = Q_{\text{net}} + Q_{\text{rch}} - Q_{\text{drn}} + Q_{\text{riv}} - Q_{\text{sub}}$$

Given the boundary conditions, such as block size, conductivity and the amount of water in initial state, a flow field is calculated in a fixed set of iterations. Basic information on the techniques is available in Hill (1990), McDonald *et al.* (1991) and Prudic (1989).

The hydrological model covers an area of approximately 24 km by 28 km. It can roughly be subdivided into three areas: a slightly raised sandy ridge running from north to south/southeast through the centre of the area and two lower areas with peat and clay soils located east and west of the sandy ridge.

The basic input of an empirical model that simulates groundwater flow is information about the initial and boundary conditions. These include (Anderson and Woesner, 1992):

- location, thickness and resistance to flow of aquitards (layers with high resistance to water flow, such as clay) and aquifers (layers with some resistance to water flow, such as sand);
- the levels and position of surface water bodies;

- locations where fixed levels of groundwater are assumed;
- locations where groundwater abstractions are made; and
- values for precipitation and evapotranspiration.

These conditions will be discussed below. Additional information is available in Heij *et al.* (1985), Heijnens (1988) and IWACO (1992, 1994).

At 200 to 250 m below the surface, a continuous, impermeable clay layer is present, which is taken as the impermeable base of the modelled system. The hydro-geological structure of the area consists of three different resisting layers of peat and clay (Aelmans, 1985; Speelman and Houtman, 1979; van der Gun, 1978). Between these layers, there are deposits of sand and gravel. The top resisting (confining) layer in the area (model layer 1) consists of marine clay and peat. Model layer 2 is a sandy aquifer. Model layer 3 is a resisting layer with glacial till, which is only present in the northeastern part of the model area. Model layer 4 is another sandy aquifer. Model layer 5 is a resisting layer formed by old river deposits, which is only present in the southwest part of the model area. Finally, model layer 6 is an aquifer underneath the hydrological base of the model.

In addition to precipitation, surface water bodies in the area are important sources of water. All surface water levels in the area are controlled. Large lakes like the IJsselmeer, the Loosdrechtse plassen, the Ankeveensche plassen and the Naardermeer are maintained with high water levels relative to the surrounding polders. The water level data are compiled from maps obtained from local water boards (updated for 1993). The Horstermeer and Bethune polders have surface water levels 1–2.5 m below the surrounding polder levels. Some of the lakes originate from peat digging, while others, like Maarsseveen and Spiegelpolder, are sandpits. The bottom of these sandpits can be 30–50 m below surface level and so reach deep into the first aquifer.

Groundwater is abstracted from a dozen locations in the area. Abstraction for drinking water started in 1888 (Witmer, 1989). Groundwater abstraction influences the flow pattern of groundwater and, therefore, provides important boundary conditions for a groundwater flow simulation. In the last 100 years, abstraction has increased and influenced the isohyps pattern (equal groundwater level lines) in the sandy ridge (Schot, 1989). Since the early 1990s, there has been a slight decrease in industrial abstraction; the total amount of water abstracted by both drinking-water and industrial companies is $42 \times 10^6\,m^3$ per year. Only those wells with an abstraction greater than $10\,000\,m^3$ in 1995 have been included in the model.

Precipitation is the primary water source for the area. Net precipitation is an important boundary condition for the groundwater flow simulation. Net precipitation is calculated by subtracting the evapo-transpiration from the gross precipitation. The evapo-transpiration depends on the structure of the vegetation and is influenced by land use. As a consequence, the spatial differentiation of net precipitation is determined by the differentiation in land use in the area.

The fixed levels that are maintained in the surface waters at the boundaries of the model area are used as boundary conditions for the simulation. These fixed levels include the river Eem at the eastern border, the Amsterdam–Rhine Canal in the west, and the IJsselmeer and Gooimeer in the north. Only the southern border of the model area is not defined by a large surface water body. Instead, a constant groundwater level in the sandy ridge serves here as a boundary condition.

Simulation experiments with the hydrological quantity model result in fluxes and heads (pressure of the water layer) for all the model elements. Water budget terms are calculated for each cell in the whole area. Model results include fluxes in groundwater flow, hydrological heads of the water tables and the quantity of water imported to and exported from the cells. To illustrate the output of the quantitative hydrological modelling, the amount of groundwater flow per grid cell is given under the Agricultural and the Nature scenarios (Fig. 7.2). Negative values indicate infiltration and positive values indicate discharge of groundwater.

7.3 The hydrological quality model

For modelling the quality of water (i.e. the chemistry in surface and groundwater), a straightforward approach is adopted. Each cell in the quantitative model of the top layer of the model is considered as a bucket. The chemistry of groundwater, surface water and precipitation in each bucket contributes proportionally to the net chemical concentration in the water bucket. For each model, cell net input or output fluxes of groundwater, surface water and precipitation are known. For the calculation of the chemistry of groundwater and surface water, just the net input fluxes are considered, since these are the only ones contributing to the water body in the model cell. Groundwater and surface water fluxes can flow in or out a model cell; precipitation is always an input flux (precipitation surplus is about 200 mm per year). If the ratio of the three contributing water sources that flow to the model cell, as well as the chemical concentrations of these three water sources, are known, the chemistry of the water that flows into a model cell can be calculated. Given the steady-state condition of the quantitative water model, and assuming full and instant mixing in both the groundwater and the surface water bodies, the chemistry of the groundwater and surface water can be calculated. The chemistry of groundwater, surface

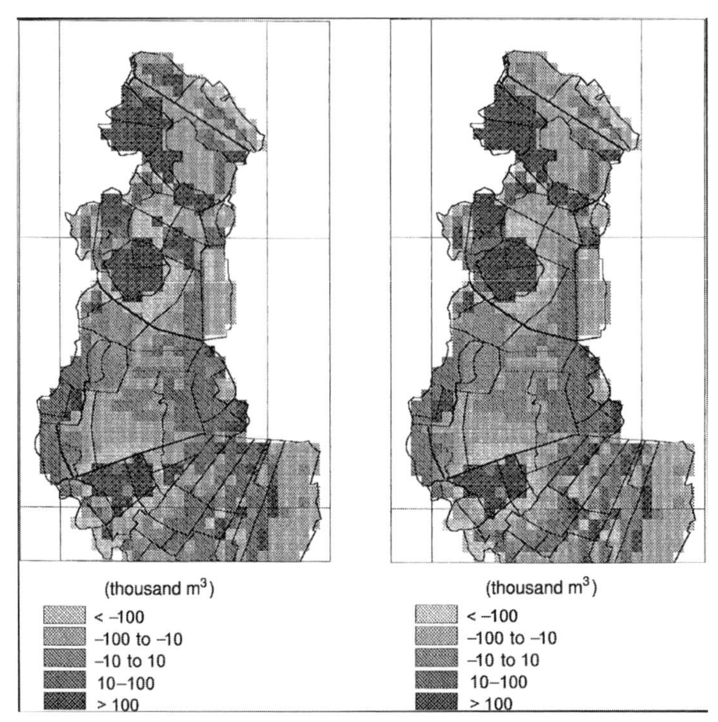

Fig. 7.2 Groundwater flow per grid cell per year (positive value is discharge, negative value is recharge) calculated under the Agriculture (left) and the Nature (right) scenarios.

water and precipitation is kept constant throughout the area and is assessed on the basis of measurements in the area (groundwater and surface water) and estimates suggested in the literature (precipitation).

Changes in boundary conditions of the quantitative model cause fluxes to model cells to change and result in revised water chemistry conditions. The changing boundary conditions are the main mechanism for the implementation of scenario changes in the model. In the context of the Nature and Agriculture scenarios, as discussed in the previous chapter, two extra model conditions are calculated with the qualitative model. Technical removal of phosphate from surface water has been implemented in the Vecht area through a phosphate-removal plant. This cleans surface water originating from the IJsselmeer, which is subsequently supplied to a large nature reserve area. Under the Nature scenario, several other polders with high nature value are assumed to be provided with such plants, which are also aimed at adapting the nitrogen levels in surface water.

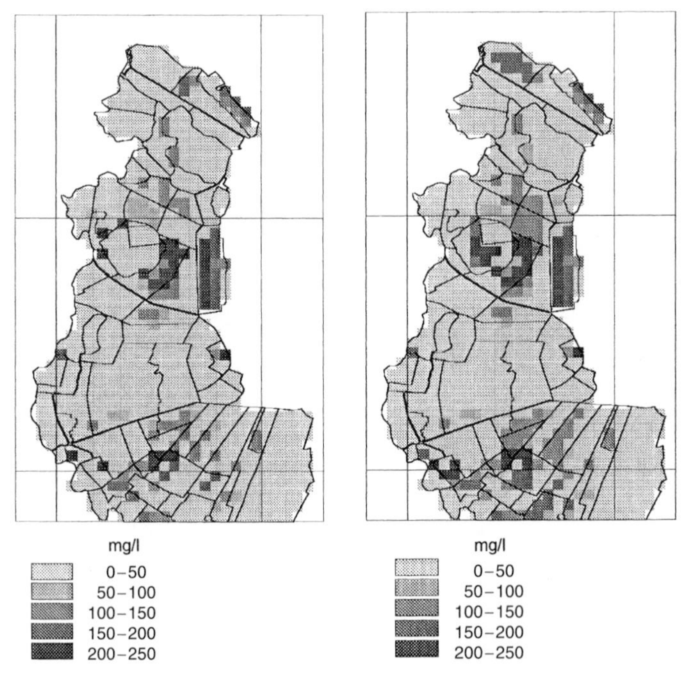

Fig. 7.3 Calculated chloride concentration in surface water per grid cell under the Agriculture (left) and the Nature (right) scenarios.

Intensifying agriculture increases nutrient levels in both surface and groundwater. This was not anticipated in the original water-quality model. The economic modelling describes the conversion of agriculture or nature under specific scenarios leading to increases or reductions in the amount of nutrient run-off in certain polders. These nutrient changes are based on field data and on the literature and are constant throughout the whole model area. The run-off at the surface with nutrients from the terrestrial areas (including an estimation of the effects of mineralising of peat soil) per scenario per polder is converted to the surface water chemistry. For the moment, this is only a rough approximation; there is insufficient information about the total volume of water in the study area. The resulting chemistry per scenario is illustrated in Figs. 7.3 and 7.4.

7.4 The ecological model

The ecological model ICHORS (Barendregt *et al.*, 1993) has been used to study the occurrence of wetland plant species. The model is based on general linear modelling regression techniques (Nelder and Wetherburn, 1974) and describes the statistical relations between the conditions in the environment and the presence of plant species. This model was selected because it can

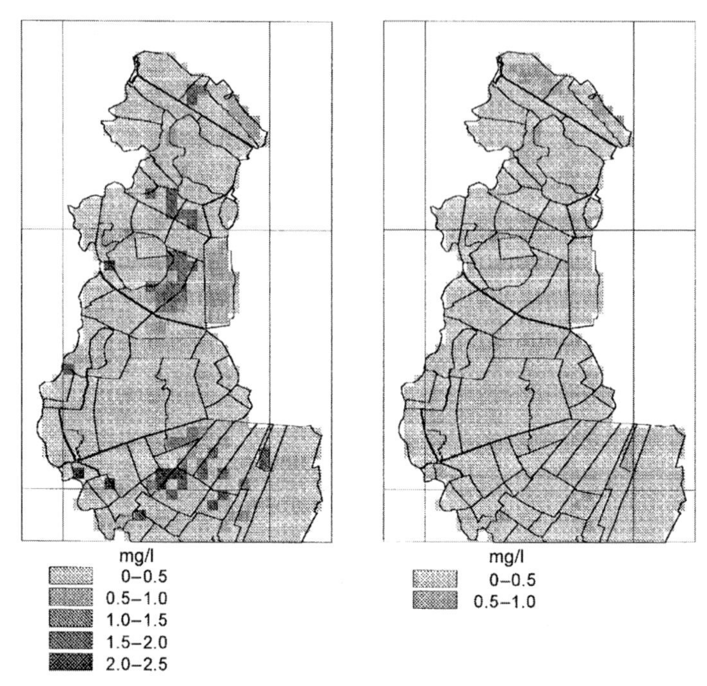

Fig. 7.4 Calculated orthophosphate concentration in surface water per grid cell under the Agriculture (left) and the Nature (right) scenarios.

be used on a grid scale with changes in environmental conditions and has been applied in earlier studies addressing changes in hydrology (see Barendregt *et al.*, 1992; Barendregt and Nieuwenhuis, 1993). Other models, based on general characteristics of vegetation or landscape, cannot be applied on the scale of grid cells. Moreover, such models, as well as models based on physiological factors, incorporate only a limited number of variables. New information on landscape variables (e.g. soil type, depth of the water) cannot be added to these models.

The statistical relations in ICHORS originate from data sets with hundreds of samples, collected in all types of wetland in the region. The samples give environmental conditions and the presence or absence of 265 plant species. The environmental conditions comprise soil type, land-use management, groundwater level, chemical concentrations of major ions and nutrients and the pH value in groundwater or in surface water. The presence of plant species is assumed to depend on a total of 25 variables.

The data are used to estimate logistic multiple regression equations that relate the presence or absence of plant species to the values of environmental variables. The procedure calculates for each variable the optimum curve of the species and selects the variable that adds the best information on the presence of the species. This procedure is continued as long as each step brings significant improvement. On average, five to six variables per plant species model are selected: these variables describe the influencing conditions of the environment around the plant in the field. The resulting regression results are stored in ICHORS. By defining values of the incorporated variables, the equations can be used to calculate the probability of encountering certain species. The model can estimate the probability of wetland plant species being found at any site with a description that fits the environmental conditions included in the estimated equations. As these conditions can result from specific management options or external events and trends, the model is suitable to calculate the effect of scenarios. Evaluation of the model output arising under different sets of condition makes it possible to assess whether a certain type of vegetation (i.e. a certain group of species) is stimulated under particular scenarios.

Two model versions are integrated in this research. ICHORS Version 2.0 incorporates relationships between locations influenced by groundwater and the presence of plant species on solid soils; ICHORS Version 3.0 includes relationships between locations influenced by surface water and aquatic and semi-aquatic plant species. Version 3.0 has been validated with independent data sets (Barendregt and Nieuwenhuis, 1993).

The model underlying ICHORS is non-spatial. In order to be used as a spatial model, the input of the model needs to consist of spatially differentiated data. The regression equations are, therefore, combined with a GIS to create a map with values for each variable. This allows the generation of a map with plant response values. An example of the results of the spatial application of the ecological model is given in Fig. 7.5.

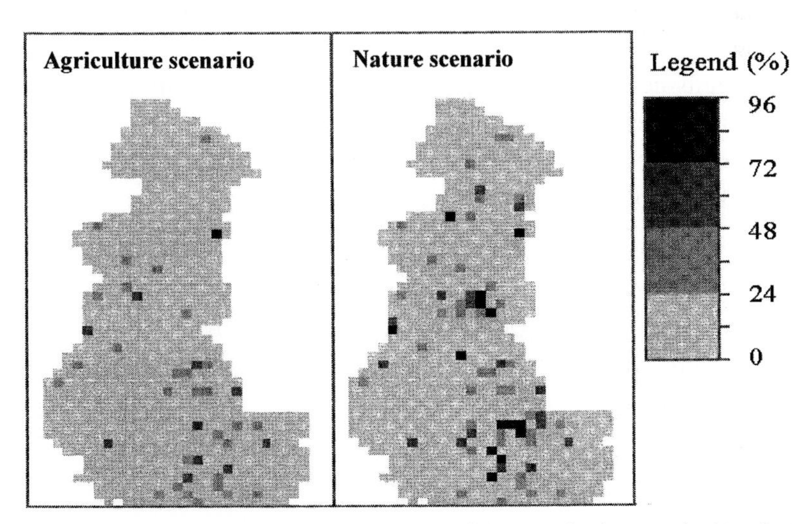

Fig. 7.5 The predicted probability of presence of the aquatic plant species *Utricularia vulgaris* per grid cell under the Agriculture (left) and the Nature (right) scenarios.

Some of the input data for the regression models is static information, for example soil texture and land use. These spatial data are taken from available maps and put in a grid format. Other data concern hydrological information on discharge or recharge of groundwater, groundwater table height and depth of surface water. These data are obtained from the qualitative hydrological model. A map of the chemical concentrations of surface water and groundwater is also calculated. At any location in the area, the water chemistry is determined by a mix of three sources of water: surface water, groundwater and precipitation. The different chemical features and proportions of the three sources determine the final chemical concentrations in the water. Processes determining concentrations in groundwater are related to groundwater flow directions and the flow time of water through the soil. Surface water chemistry is influenced by the origin of surface water, for example Rhine river water, pollution sources upstream, groundwater and precipitation. The calculations are summarised in Section 7.5.

7.5 Modelling results

The three models are applied in one computerised sequence, including the calculation of water levels and resulting flows in groundwater and surface water, the chemical composition of water, and the ecological predictions of species. In this way, for each cell in each scenario, the probabilities of encountering the 265 plant species are calculated, given the defined input by the hydrological (surface and groundwater), chemical and management (also nutrient input) conditions arising from the scenarios. Consequently, the ecological response to variables driving the regional hydrology can be estimated. Water-flow results can be aggregated to polders, which are a useful spatial level to study the interaction between hydrology, economics and policy.

The best environmental conditions to stimulate characteristic wetland species in the Vecht area are:

- high water tables resulting in very wet conditions;
- fresh conditions, so that the water has low concentrations of chloride, magnesium, sulphate, sodium, potassium, etc.;
- nutrient-poor conditions, i.e. very limited concentrations of phosphorus, nitrogen, ammonia and potassium; and
- buffered conditions, i.e. high pH values from high concentrations of calcium and hydrocarbonate.

The modelling results show two opposing influences on water quality and, hence, the presence of fen and peatland species. First, because of lower surface water tables, drainage in the peatland regions of the Vecht area results in lower groundwater tables, with the consequence that groundwater from surrounding areas with higher groundwater levels (the hill ridge) will discharge into the drained areas. This discharged groundwater has a positive effect on the wetland species as it is fresh, relatively nutrient poor and rich in calcium and hydrocarbonate. Somewhat surprisingly, drainage of a wetland positively influences the ecosystem through water chemistry effects.

Second, maintaining high water tables in polders with nature that are adjacent to polders where drainage occurs results in groundwater flows from the high to the low level. Where polders have a high water level as a result of infiltration of groundwater, their water levels will fall. To maintain the high water table in a polder where water is constantly disappearing, water has to be supplied. This water must originate from the river Vecht, which is rich in nutrients and has a relatively high salinity; this will have an adverse effect on wetland species. Moreover, the areas with high water tables are sensitive to acidification when the input of rainwater is dominating.

Thus, management aimed at stimulating nature can have unintended effects. Management intended to stimulate conditions for agriculture implies a type of water management opposite to that appropriate for nature areas. However, the effects of such management will be counterbalanced by water chemistry processes. In addition, agricultural management leading to drainage of the peatlands results in the mineralisation of peat and, the intensification of agriculture means extra input of manure and fertilizers. Both imply increases in the nutrient concentrations in the water. As a result, it would be desirable to stimulate nature only in certain parts of the region: namely, where all four environmental conditions mentioned above – wet, fresh, nutrient-poor and buffered – are satisfied.

Is it feasible to model an ecosystem of a wetland area with all its diverse characteristics? It would appear that the knowledge and tools are available to perform this task. Groundwater levels and flows can be calculated for a large area with a level of detail that makes it possible to address local variations. Surface water levels are fixed in the polders by human management. Data on the chemistry of groundwater and surface water are available and can be combined. An ecological description of the occurrence of plant species can be based on the abiotic conditions, some of which are amenable to control by management. The integrated modelling with an explicit spatial dimension makes it possible to combine the scale of regional groundwater flow with the local conditions that plant species experience.

Various improvements are still possible. The flow of surface water in the polder area is not known precisely. In the modelling, we split this flow into two parts, based on an area with a direct input of supply water and another area close to the hill ridge with a mixed input of supply water and groundwater. The chemical modelling was kept simple by means of conservative mixing. In reality, the equilibrium of the major ions and nutrients is influenced by numerous conditions such as acidity, oxygen concentrations, the buffer capacity of soil and the retention of ions; all will influence the concentrations at each spot. The modelling provides information about the general condition that prevails and not about the specific conditions, which may vary per square metre. Of course, significant variation within a grid cell is possible in reality.

Nevertheless, the results of the hydro-ecological modelling make sense. Plant species characteristic for fresh and nutrient-poor locations are predicted at isolated locations with a discharge of groundwater. Many species characteristic of polluted water are predicted at locations where river water dominates the water balance. It would appear, however, that the occurrence of many species is only influenced in the grid cells that are very affected by the conditions of the scenarios. Although the local population of such species will respond, changes in the total population may be less significant in the evaluation for the whole area. These issues will be discussed later, in Section 9.2 on indicators.

8

The spatial–economic model: agriculture, nature conservation and outdoor recreation

8.1 Introduction

This chapter looks at the changes in costs, benefits and nutrient flows under the three scenarios, as formulated in Chapter 6. The goals of this chapter are (a) to present the economic indicators, calculate these under each scenario and perform a financial cost–benefit analysis (CBA), and (b) to calculate the changes in the run-off of nutrient flows under each scenario. The economic cost and benefit indicators will be entered, along with the ecological and spatial equity indicators (Ch. 9), into the scenario evaluation (Ch. 10). Additional economic indicators, to be calculated, are the changes in net present value (NPV) and employment, and the effects of changes in the spatial and economic structure on nutrient flows. The last leads to the calculation of environmental indicators for run-off and surplus of nitrogen and phosphate, which serve as inputs to the ecosystem model discussed in Chapter 7.

There are several types of CBA. In this chapter, financial CBA indicators are generated, which means that only financial transactions are considered. Another approach is economic CBA, which also includes (social) costs and benefits for non-market goods and services, such as those related to changes in nature (Zerbe and Dively, 1994). As no prices are available for these non-market valuations, values must be estimated. The economic CBA would perhaps seem more relevant for the purpose of the present study, as the benefits and costs of improved environmental quality and nature in a specific scenario fall outside the realm of the market. Instead, however, this study will use ecological indicators to measure the improvements in environmental quality. The economic and ecological indicators will then be combined in the evaluation. If the environmental improvements were also measured through economic indicators in the evaluation, the problem

Table 8.1 *Input and output of the spatial–economic model*

Input	Output
Scenarios and settings (per polder, including the number of hectares)	Economic indicators (per polder)
Economic and nutrient data (per hectare)	Nutrient indicators (per polder)
Environmental quality indicators (per polder)	

of double-counting would arise. Therefore, the economic model focuses attention on financial CBA and indicators. A financial CBA is performed for each of the three scenarios. In addition, an economic CBA is performed for the Recreation scenario to illustrate the outcomes of an evaluation based on economic indicators that take account of environmental improvements. The results of this will not be used in the final evaluation later on.

In order to perform the CBA, a spatial–economic model is formulated. This model describes mainly agriculture, nature conservation and outdoor recreation. Table 8.1 lists the inputs and outputs of the spatial–economic model. The inputs to the spatial–economic model consist of the scenarios and settings (as defined in Ch. 6), economic and agricultural–environmental data at the hectare level for the various scenarios and settings (calculated in this chapter) and the environmental quality indicators per polder (which will be calculated in Ch. 9). The output generated by the economic model includes environmental indicators, which enter the hydrological model, and economic indicators, which enter the evaluation.

The economic and nutrient indicators are calculated per hectare for each setting in the scenarios. Subsequently, the changes in land use are calculated at polder level. The indicators are then calculated for changes, relative to the Reference scenario, for the Agriculture, Nature and Recreation scenarios. Under the Agriculture scenario, agriculture is intensified in various polders (see polder settings in Table 6.1). The revenues are calculated per hectare in the present and in the intensified situation. The changes in revenues compared with the present situation are the benefits. In addition, employment and environmental indicators are calculated. Under the Nature scenario, parts of the agricultural land are converted into nature. The costs of this conversion include those of acquisition, restructuring and maintenance, as well as the costs of foregone benefits of agriculture (i.e. the opportunity costs). These opportunity costs are the current revenues of agriculture. The benefits of conversion to nature are not calculated (see above). The environmental indicators that change through a reduction in agricultural land are measured in physical–biological terms. The

Recreation scenario builds upon and extends the Nature scenario: the land areas that are converted into nature are opened for recreation; some polders that are open water are opened to recreation; and some polders are flooded for the purpose of water-based recreation.

The spatial–economic model is applied to the three scenarios. The treatment of scenarios is mutually consistent from an economic perspective, since the benefits per hectare of agricultural land serve as opportunity costs in the Nature and Recreation scenarios.

This chapter is organised as follows. Section 8.2 discusses the data on which the calculations are based. Section 8.3 presents the model indicator results for the Agriculture, Nature and Recreation scenarios. Section 8.4 shows the type of result that arise from an economic CBA of the Recreation scenario.

8.2 Environmental and economic data

8.2.1 *Agriculture*

Agriculture is an important activity in the Vecht area. More than half of the land in the area is dedicated to agriculture, the most important type being dairy farms. This section describes the data that will be used in the model part on agriculture. The model distinguishes between two types of agriculture: present situation and intensive agriculture. The data are for 1994 and 1995.[1]

Present situation

The present situation is considered to be the regional average for the province of North-Holland. Although the Vecht area is situated in the provinces of Utrecht and North-Holland, the agricultural intensity of North-Holland represents the Vecht area better. The reason is that the data for Utrecht are less representative as they include an intensive agricultural area outside the Vecht area. The first column of Table 8.2 shows the data for the present situation. The detailed model used in the calculation is presented in Appendix 8.1. The number of dairy cows per hectare is derived from the Dutch Central Bureau of Statistics (CBS) data and is valid for the year 1995 (CBS, 1998). The size of farms, and economic and labour parameters, are estimated on the basis of the data from the Landbouw Economisch Instituut (LEI, 1998). The economic data that are valid for each type of agriculture is the milk price per kilogram (LEI, 1998; Coberco, telephone conversation 1998).

[1] Vreke and Veeneklaas (1997) have performed an economic CBA for the Netherlands (*Economic Main Structure*) that partly covers the Vecht area. Their study includes detailed information on different types of nature and agricultural activity.

Table 8.2 *Data on types of agriculture*

Parameters related to	Parameters[a]	Present agriculture	Intensive agriculture
Economy	Total revenue (€)	2354.21	5507.08
	Dairy cows (per ha)	0.8	1.74
	Employment (per ha)	0.1002	0.0914
Environment	Nitrogen surplus (kg/ha)	241	612.5
	Phosphate surplus (kg/ha)	44	110.25
	Run-off nitrogen (kg/ha)	37.13	94.36
	Run-off phosphate (kg/ha)	6.78	16.98

Sources: CBS (1998), Daatselaar *et al.* (1990), LEI (1998), Provincie Noord-Holland (1997).
[a] Based on area of agricultural land.

The environmental data are as follows. The phosphate surplus as a percentage of the nitrogen surplus is 18%, and the run-off as a percentage of the surplus is 15.41% (Provincie Noord-Holland, 1997). The run-off as a percentage of the surplus is estimated by averaging the nitrogen run-off on clay and sand (Antuma *et al.*, 1994). No separate treatment was followed for different soil types because the run-off mechanism in peat soils is complex and insufficiently understood. The part of total revenue per hectare that remains as family income per hectare is 15.83% (LEI, 1998). Because of the lack of detailed data, it was impossible to derive complete nutrient balances for the present situation.

Intensive agriculture

Agriculture is intensified by increasing the number of cows per hectare. Table 8.2 shows the different parameters that are used for intensive agriculture. The economic, farm and labour parameters for this type of agriculture are derived from LEI (1998, Tables 3.04, 3.05 and 3.06). The environmental parameters are taken from Daatselaar *et al.*, (1990). The use of fertilizer, enriched-fodder and atmospheric deposition determine the inflow of nitrogen. The outflow of nitrogen occurs via milk and meat. The calculation of the total surplus of nitrogen includes an allowance for the evaporation of ammonia. As it was impossible to construct a nutrient balance for phosphate, the surplus of phosphate was estimated to be 18% of the surplus of nitrogen.

8.2.2 *Nature conservation*

The goal of this section is to assess the costs of converting a part of the Vecht area from agriculture into nature. These costs will be specified at the polder level.

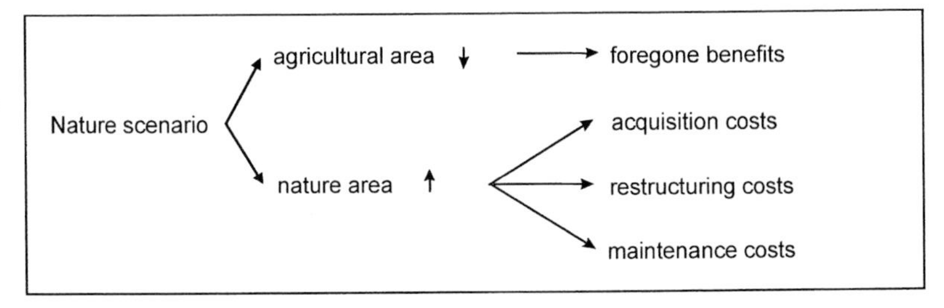

Fig. 8.1 Schematic view of the various costs related to the Nature scenario.

Costs of stimulating nature

By converting land from agriculture to nature, various types of cost arise. The first type are opportunity costs (or foregone benefits) resulting from a reduction in the agricultural area. The second type of cost is related to an increase of area for nature. The costs of stimulating nature are divided into four categories (Fig. 8.1): acquiring costs, restructuring costs, maintenance costs and opportunity costs (foregone agricultural benefits). The costs of acquiring land arise at one point in time. The costs of restructuring are spread over 10 years, and it is assumed that the annual cost during the first 10 years is a tenth of the total restructuring costs. The maintenance and opportunity costs are yearly costs. The costs of restoration and conservation in the Vecht area that do not depend on the Nature scenario have been excluded. An example is the costs of restoration of the river Vecht, since this action is performed under all scenarios and, therefore, does not give rise to differences among the scenarios.

The acquisition, restructuring, maintenance and opportunity (foregone benefits) costs are calculated as follows.

- The costs of acquiring land in the North-Holland part of the Vecht area is €22 681 per hectare (in 1997 prices) (Provincie Noord-Holland, 1995). It is assumed that these costs are the same for the Utrecht part.
- For restructuring the North-Holland part of the Vecht area, the costs are estimated to be €10 181 per hectare (in 1997 prices) (Provincie Noord-Holland, 1995). Also here, the costs for the Utrecht part of the Vecht area are assumed to be equal to those of the North-Holland part.
- The maintenance costs are estimated per hectare of nature. These depend on the type of nature and the way nature is managed. Based on the maintenance costs of the existing nature areas that are the property of Natuurmonumenten and Staatsbosbeheer, the average maintenance costs per hectare are estimated to be €470.3 per year (Bos and van den Bergh, 2002). A more refined estimate of these costs could be obtained by classifying the type of nature management in each polder.

Table 8.3 *The capacity and capital costs of phosphate-removal plants*

Polder	Average capacity (m^3/h)	Maximum capacity (m^3/h)	Amount of water treated (m^3/year)	Installation costs (€/m^3)	Capital costs (€/m^3 per year)[a]	Net present value costs (5% interest 10 years) (thousand €)
Naardermeer	300	600	500	0.0644	0.120	494
Kortenhoef	1000	2000	5000	0.0545	0.0971	4008
Noorderpark	1000	2000	10000	0.0545	0.0971	8017

[a] The capital costs per year are for a plant extended to clean class 2 mud.

- By converting agricultural land into nature, the benefits of agriculture are foregone. These are opportunity costs, the level of which depends on the type of agriculture (normal, intensive or extensive) that was present before the land was converted into nature. The opportunity costs of 1 ha per year of 'normal' agriculture is €2354.21 and of 'intensive' agriculture is €5507.08 (Section 8.2.1).

Phosphate removal

Under the Nature scenario, phosphate-removal plants are installed. These installations remove phosphorus from the water that flows into the Vecht area. These plants differ in their rate of phosphate removal. In the Vecht area, the plants may be installed in the Naardermeer, west of Kortenhoef and south of Noorderpark. Table 8.3 shows the capacity and the costs of those plants (Stowa, 1994). At Kortenhoef and Noorderpark, the same type of plants may be installed. The capacity and costs are given per cubic metre of treated water.

8.2.3 *Outdoor recreation*

The financial benefits of outdoor recreation can be estimated by using two types of data: spending on recreation and the revenues of recreation-based industries. The costs of the scenario are derived from the data for nature, except for the restructuring costs to convert agricultural land into recreation areas.

Benefits of outdoor recreation based on spending on recreation

The CBS distinguishes three types of recreation: day trips, short holidays and long holidays (CBS, 1996b). A day trip is defined as (a combination of) activities, such as hiking, biking and fishing, with a duration of at least two hours and starting from the home address. Short and long holidays take place at an address other than the home address. The duration of a short holiday is on average 3.1 days and that of a long holiday on average 10.4 days. Expenditures for each category are as follows.

Day trips The amount of money that is spent on day trips is calculated from statistical data. The precise number of day trips in the Vecht area is not known but is approximated on the basis of the number of day trips by inhabitants of the region, the average number of day trips per person per year, and the average amount of money spent on a day trip (CBS, 1996c, 1997b). The total amount of money spent on day trips in the Vecht area each year is estimated to be €244.59 million.

Short and long holidays No accurate and complete data are available on the number of short and long holidays in the Vecht area. These numbers are estimated from data on holidays in the 'Utrecht-Holland lake area' and 'Utrechtse Heuvelrug and Het Gooi' (CBS, 1996b), and from the average expenditure per short holiday (CBS, 1997b). The total expenditure on short holidays is estimated to be €11.28 million, which is €1850 per hectare per year. The number of long holidays in the Vecht area is calculated in the same way as the number of short holidays (see Appendix 8.2), and the total expenditure on long holidays is estimated to be €21.33 million, which is €3525 per hectare per year.

Benefits of outdoor recreation based on total revenues of recreation-related sectors

An estimate of the total sales (i.e. cash flow) in the Vecht area is based on the number of recreation companies and their sales. The recreation sectors include hotels, restaurants, bars, discos, sports centres (including sail and surf schools), exploitation of recreational vehicles with crew, and yacht havens. The total sales for recreation are estimated to be €360.5 million per year (Bos and van den Bergh, 2002). The employment in recreation in the Vecht area is estimated at 5928 people, which is about one person per hectare of recreation (Bos and van den Bergh, 2002).

Summary of financial benefits of outdoor recreation

Table 8.4 presents the benefits of outdoor recreation, which are estimated on the basis of (a) the spending of people that recreate and (b) the revenue

Table 8.4 *Total benefits in the Vecht area*

	Total benefits (million €)	Total benefits per hectare of recreation (million €)
Total spending on recreation	277	45.45
Total revenues of recreation-based sectors	360.5	59.1

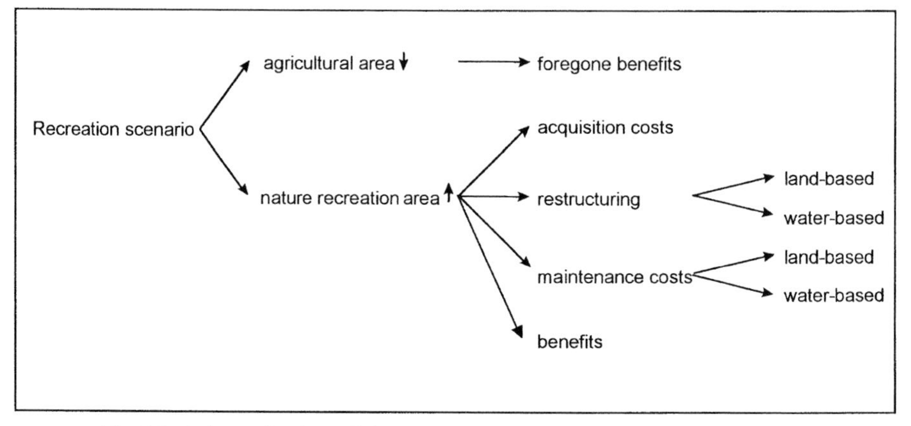

Fig. 8.2 Schematic view of the various costs related to the Recreation scenario.

of recreation-based sectors. In the subsequent analysis, the result of the former approach is used. The latter approach probably gives rise to an overestimation because not all revenues of recreation-related sectors are from outdoor recreation. For instance, visits to restaurants are also made by people living nearby and who are not participating in outdoor recreation.

Costs of stimulating recreation

In the stimulation of recreation scenario, parts of certain polders are converted from agricultural land to recreation areas. Various types of cost are associated with this land conversion. Figure 8.2 presents these schematically. First, agricultural land will be converted into recreation areas. The associated acquiring costs have to be paid once and are equal to €20 866 per hectare (Provincie Noord-Holland, 1995). Second, the land has to be restructured for land- or water-based recreation. These costs are estimated to be €15 882 per hectare (Provincie Noord-Holland, 1995). The costs of restructuring the agricultural land into open water are assumed to be €4538 per hectare (estimation based on Provincie Noord-Holland, 1995 p. 23). The acquiring and restructuring costs need to be paid at one point in time.

In addition, there are yearly maintenance and opportunity costs. The maintenance costs for land-based recreation are assumed to be 10% higher than the costs of maintaining nature: €517.3 per hectare per year. The maintenance costs for water-based recreation are 50% of those of land-based recreation: €258.7 per hectare per year (own estimate). The opportunity costs are equal to the benefits that are foregone as a result of the conversion of agricultural land. The present type of agriculture has a yearly economic benefit of €2354.21 per hectare. These costs are taken as the opportunity costs.

As under the Nature scenario, the costs of restoration and conservation that are planned independently of the Nature or Recreation scenario are excluded because they do not create any differences among scenarios. Finally, as with the Nature scenario, the Recreation scenario includes the installation of phosphate-removal plants.

8.3 Modelling results under the three scenarios

8.3.1 *Stimulation of agriculture*

This section presents the results of the Agriculture scenario. The settings of each polder were presented in Table 6.1. The economic settings are: 0, no changes; 1, half intensification (i.e. 50% of the agricultural land to be intensified); and 2, full intensification (i.e. 100% of the agricultural land to be intensified). Intensified use of agricultural land results in more revenues, but also more nutrients.

Table 8.5 shows the changes in the economic output variables per region compared with the present situation.[2] Although the analysis is performed at the polder level, the results for each of the 73 polders are aggregated into three regions to present a concise overview of the economic and environmental indicators. In Appendix 8.3, the results are presented per polder level for each scenario. The NPV is calculated with a 5% interest rate and over a time period of 10 years. The environmental indicators are presented in Table 8.6 for the various regions.

8.3.2 *Stimulation of nature*

In the Nature scenario, some parts of the Vecht area are converted from agriculture to nature in order to stimulate the ecological functions of the polders

[2] The three regions are defined as follows: the north contains polders 1–25, 27, 28, 73 (north of the Hilversums Kanaal); the middle contains polders 26, 29–38, 41, 46–49, and 51 (between the Hilversums Kanaal and the Tienhovens Kanaal); and the south contains polders 39, 40, 42–45, 50 and 52–72 (south of the Tienhovens Kanaal).

Table 8.5 *Economic output per region under the Agriculture scenario (changes relative to the present situation)*

Region	Agricultural base (ha)	Agricultural land change (ha)	Total revenue (million €)	Employment (No. persons)	Dairy cows (No.)	Net present value (5% interest, 10 years) (million €)
North	4500	3163	9.97	−27.83	2973	76.99
Middle	1775	1438	4.53	−12.65	1351	35.00
South	3550	2613	8.24	−22.99	2456	63.60
Total	9825	7214	22.74	−63.47	6780	175.59

Table 8.6 *Environmental output per region under the Agriculture scenario (changes relative to the present situation)*

Region	Agricultural base (ha)	Agricultural change (ha)	Surplus nitrogen (million kg)	Surplus phosphate (thousand kg)	Run-off nitrogen (thousand kg)	Run-off phosphate (thousand kg)
North	4500	3163	1.26	209.5	181.0	32.3
Middle	1775	1438	0.57	95.2	82.3	14.7
South	3550	2613	1.04	173.1	149.5	26.6
Total	9825	7214	2.87	477.8	412.8	73.6

in this region. The choice of agricultural areas that can be converted into nature is very site specific. There are three settings in the Nature scenario. Under setting 0, the present land use and the water level do not change. Under settings 1 and 2, the water level is increased by 0.1 or 0.2 m, respectively, and 50% or 100%, respectively, of agricultural land is converted into nature areas.

For these specific choices, certain goals and constraints were taken into account: a goal was to obtain a ‘continuous’ nature area, for example in the northeast region around the Naardermeer (polder 6); a constraint was that the western part of the area should remain open grassland, with the exception of parts where forests have been developed. For agriculture and recreation, other goals and constraints apply. Therefore, given the goals and constraints in mind, each polder was analysed as to whether the agricultural area (or a part of it) could be transformed into nature.

The decision as to which parts of the polders are converted depends on environmental, physical, spatial planning and hydrological aspects. Around the major lakes, the agricultural areas are completely converted into nature, while in a north–south corridor from the Naardermeer (polder 6) to the Loosdrechtse plassen (polder 35), part of the agricultural area is converted into nature.

Table 8.7 presents the aggregated estimated costs for the north, middle and south regions, in which (a part of) the agricultural area is converted into nature areas. This table also includes the NPV of all costs based on a discount rate of 5% and a time horizon of 10 years.

Two types of cost are paid at a single point in time: the costs of acquiring agricultural land (€86.8 million) and the restructuring costs (€42.3 million) (see Table 8.7). Maintenance costs of €1.96 million are expended yearly. The yearly opportunity costs are €9.80 million.

The spatial distribution of these costs over the polders allows decision makers to choose specific polders for nature creation. For example, they may decide to convert the agricultural areas of the five polders (polders 7, 9, 10, 14, 18) around the Naardermeer (polder 6) into nature areas, and leave the other polders unchanged.

The benefits of this scenario cannot be calculated directly, because there is no market price for nature (i.e. no financial benefits). To estimate the benefits of nature stimulation in monetary terms, there are various valuation methods to assess the non-market values associated with environmental changes. As discussed in the introduction to this chapter, no valuation is performed for the Nature scenario, because in the evaluation the 'benefits' are represented by changes in the ecological indicators. Nevertheless, in Section 8.4, a separate analysis including non-market values will be performed for illustrative purposes, as well as for comparison with the evaluation approach in Chapter 10.

The environmental indicators are presented in Table 8.8. These nutrient and phosphate indicators are directly derived from the reduction in the number of hectares of agricultural land.

8.3.3 *Stimulation of recreation*

This section describes the Recreation scenario. The province of North-Holland has adopted a plan to create a recreation area of about 400 ha, of which 280 ha will be in the Vecht area. The recreation area has the most favourable impact in terms of environmental, nature conservation and economic objectives when (a) recreation occurs as close as possible to Amsterdam, (b) the character of the Vecht area as an open area remains (landscape value), and (c) the agricultural interests in terms of a connected agricultural area remain intact (Provincie Noord-Holland, 1995). The central goal of outdoor recreation is to develop and increase the recreational function of the Vecht area. Therefore, a number of actions are necessary: (a) improve the (external) infrastructure to facilitate access to the region, (b) improve the internal infrastructure by developing safe cycling

Table 8.7 *The economic indicators per region under the Nature scenario*

Region	Agricultural land change (ha) (Reference)	Acquiring costs (million €)	Restructuring costs (million €)	Maintenance costs (thousand €)	Opportunity costs (million €)	Employment (No. people)	Net present value (million €)
North	2050 (4500)	42.78	20.87	964.2	4.83	−205.41	−109.02
Middle	650 (1775)	13.56	6.62	305.7	1.53	−65.13	−34.20
South	1462.5 (3550)	30.52	14.89	687.9	3.44	−146.54	−83.16
Total	4162.5 (9825)	86.86	42.38	1957.8	9.80	−417.08	−226.38

Table 8.8 *Environmental indicators per region under the Nature scenario*

Region	Agricultural land converted (ha) (Reference)	Surplus nitrogen (thousand kg)	Surplus phosphate (thousand kg)	Run-off nitrogen (thousand kg)	Run-off phosphate (thousand kg)
North	2050 (4500)	−494.05	−90.20	−76.12	−13.90
Middle	650 (1775)	−156.65	−28.60	−24.14	−4.41
South	1462.5 (3550)	−352.46	−64.35	−54.30	−9.92
Total	4162.5 (9825)	−1003.16	−183.15	−154.56	−28.23

and hiking tracks and canoeing ways, and (c) provide sufficient clean and open water for fishing. Under the Recreation scenario, land- and water-based recreation are stimulated. Settings 0, 1 and 2 are equal to those in the Nature scenario, except that nature is now also opened for land recreation. In setting 3, two polders are flooded for water recreation, and in setting 4 water recreation is stimulated in polders with open water.

Table 8.9 gives the financial costs and benefits of the Recreation scenario per region. The benefits are calculated as follows. The total spending on recreation (per hectare open for recreation) is estimated to be €45.45 million (see Table 8.4). For every hectare that will be opened for recreation, the benefits are assumed to change proportionally.

The acquisition and restructuring costs are paid once, and the maintenance and opportunity costs occur regularly over time. The benefits occur regularly over time as well. The spatial distribution of the results allows decision makers to choose specific polders for nature–recreation stimulation.

The last column of Table 8.9 presents the NPV of the benefits minus the costs. This is calculated with a discount rate of 5% per year and a time horizon of 10 years. The positive value shows that recreation has a positive value in financial terms. Note that the costs are paid for by the public sector (municipality, province or central government) while the benefits are earned by the private sector. The values of the nutrient indicators under the Recreation scenario equal those under the Nature scenario, because the agricultural land that is converted into nature and recreation is equal under both scenarios. The costs of the phosphate-removal installations are equal to those under the Nature scenario discussed above.

In Chapter 9, an environmental quality indicator will be presented. This indicator is used here to illustrate the possible feedback from biodiversity and nature quality to outdoor recreation. This reflects the idea that more people will visit the Vecht area when its environmental quality is higher. This implies

Table 8.9 *The costs and benefits of converting land from use by agriculture into use by nature–recreation*

Region	Land converted (ha) (Reference)	Open water for recreation (ha)	Acquiring costs (million €)	Restructuring costs (million €)	Maintenance costs (million €)	Opportunity costs (million €)	Benefits (million €)	Employment (No. people)	Net present value (5% discount rate, 10 years) (million €)
North	2050 (4500)	275	42.78	29.32	0.99	4.83	112.49	1980	698.73
Middle	650 (1775)	1600	13.56	17.58	0.75	1.53	102.27	585	741.61
South	1462.5 (3550)	350	30.52	24.82	0.85	3.44	82.38	1316	542.29
Total	4162.5 (9825)	2225	86.86	71.72	2.59	9.80	297.14	3881	1982.63

Table 8.10 *The environmental quality indicators under the Reference and Recreation scenarios*

Region	Reference	Recreation	Difference
North	1152	1240	89
Middle	776	809	33
South	1196	1288	93
Total	3124	3337	215

Table 8.11 *Financial benefits and the net present value under the Recreation scenario including the environmental quality indicator per region*

Region	Agricultural land converted (ha)	Open water converted (ha)	Benefits (million €)	Net present value (million €)
North	2050	275	132.13	905.96
Middle	650	1600	110.74	807.01
South	1462.5	350	106.45	728.10
Total	4162.5	2225	349.32	2441.07

that higher environmental quality increases the benefits per hectare of land open to recreation. Table 8.10 presents the environmental quality indicators for the Reference and the Recreation scenarios. The environmental quality indicator on the polder level takes values of between 0 and 100, which are then aggregated to the regional levels.

Table 8.11 gives the financial benefits and the NPV of the Recreation scenario when the changes in the environmental quality indicator are taken into account and proportional benefit changes are assumed. This table can be compared with Table 8.9 in which the benefits and NPV of the Recreation scenario are presented without considering the effect of changes in environmental quality. Note that the results in Table 8.11 are indicative, because the effect of the environmental quality indicator on the benefits of recreation is not measured. It is assumed that an increase in the indicator by 1 point increases the benefits by 5%.

8.4 Economic benefits of nature and recreation

This section shows the results of an economic CBA of the Recreation scenario. These results are presented only for illustrative purposes. It should be emphasised that they are not used in the final evaluation procedure. If the economic benefits and the ecological gains were to be taken into account in the evaluation, there would be a double-counting problem.

The question here is how to value recreation in the absence of market prices for recreational services. In this study, the goal is to estimate the change in value of an increase in the number of recreational sites and, if possible, a change in the quality of recreational sites (e.g. by using a quality indicator). For an economic analysis, it is important to know the willingness-to-pay (WTP) to visit the recreational site. This WTP may be measured for recreation through various methods, such as the travel cost method (TCM) and the contingent valuation method (CVM). CVM studies of wetland recreation have been mainly performed for the USA and the UK (see Klein and Bateman, 1998). CVM requires extensive field research and questionnaires for a specific site. This is very costly, in terms of both money and time.

In many cases, it is practically or financially impossible to perform a TCM or a CVM. Then value transfer can be used (see Section 2.3.2). This method applies monetary values from a particular valuation study to another study, often in another geographical area than the one where the original study was performed. The method of value transfer is mainly used to value non-marketed resources, for example damage to natural resources (Parsons and Kealy, 1994), or recreational demand (McConnell, 1992). Brouwer and Spaninks (1999) test the validity of environmental value transfer. Problems related to value transfer for water-quality estimates are analysed in Desvousges *et al.* (1992).

Few valuation studies focusing on wetlands are available for the Netherlands. Ruijgrok (1998) has performed a CVM to estimate the WTP for various types of nature in the coastal zones in the Netherlands. Use and non-use values are derived for various nature types. The use values range from €0.65 to 2.54 per visit, depending on the nature type. The average non-use value is €5.13 per year. When corrected for protest- and zero-bidders, the average non-use value is €15.17 per year. Corrected for extreme high bidders, the average non-use value is €4.12 per year.

De Groot *et al.* (1998) have estimated the economic value of the Oostvaarder-splassen (a nature area in the centre of the Netherlands). A questionnaire gave the following results: a use value of €179.02, a non-use value of €97.65 and a total value of €276.67 (all per person per year).

Brouwer *et al.* (1999) performed a meta-analysis of wetland CVM studies for different wetland characteristics. The average WTP is calculated for four wetland functions. The average WTPs for three use values are €126.06 for flood control, €29.27 for water generation and €71.47 for water quality. For the use value of biodiversity, the average WTP is €103.60. The authors state that these numbers

Table 8.12 *Net economic value of various recreation activities*

Activity	Net economic value (1987 US$/recreation day)
Hiking	29.08
Sunbathing/swimming	22.97
Cold water fishing	30.62
Motorised boating	31.56
Total	114.23

Source: Walsh *et al.* (1992).

need to be taken with caution because CVM design has a significant impact on the estimated WTP.

Various studies have tried to estimate the value of recreation activities. Such studies have mainly been performed in the USA and the UK. Walsh *et al.* (1992) performed a meta-analysis of various studies on outdoor recreation in different geographical areas. The basic data of the analysis are the net economic values per recreation day from several TCM and CVM studies undertaken between 1968 and 1988: 'The net economic values are equivalent to the dollar (monetary) amount participants would be willing to pay over or above their current expenditures to ensure continued availability of the opportunity to use recreation resources' (Walsh *et al.*, 1992). Table 8.12 gives the values for the recreational activities that are relevant to the Vecht area.

The studies that are analysed in Walsh *et al.* (1992) include various activities that cannot be performed in the Vecht area (or elsewhere in the Netherlands) for example big game hunting and visits to isolated wilderness. Interestingly, the activities with the highest net economic value per recreation day are certain types of fishing and game hunting. Various studies generate indicators for the value of other recreation activities. Pearson (1992) estimates the recreational and environmental value of Rutland Water (a large reservoir in England). The average WTP in pounds sterling in 1992 (£1 is approximately 2003 €1.47) was £13.11 for biking, £24.77 for canoeing/rowing and £23.61 for sailing/wind-surfing. Creel and Loomis (1992) estimate the recreation value of water for wetlands in the San Joaquin Valley (USA) and derive the following WTP per person per year (use benefit estimates): approximately $140 for one of the activities viewing, fishing or hunting, approximately $420 for two activities and $562 for all three activities (1992 US$1 is approximately equivalent to €0.92 in 2003). The model shows that the use benefit estimates increase by 10–30% when the water quantity is

Table 8.13 *The economic benefits of converting agricultural land into nature–recreation areas*

Region	Agricultural land converted (ha)	Net economic value (million €)	Use value (million €)	Non-use value (million €)
North	2050	65.11	3.26	8.43
Middle	650	20.64	1.03	2.68
South	1462.5	46.45	2.33	6.03
Total	4162.5	132.20	6.62	17.14

increased. Other studies on (wetland) nature valuation are by Loomis (1987) and Cordell and Bergstrom (1993).

It is difficult to transfer these benefits to the Vecht area because these values pertain to geographical areas of quite a different size to the small Vecht area. For the benefit transfer of the Vecht area, the use and non-use values of nature obtained by Ruijgrok (1998) will be used, because they are more comparable with the Vecht area. Ruijgrok (1998) estimates the values of nature in coastal zones in the Netherlands as a use value of €1.59 per visit and a non-use value of €4.12 per year. Table 8.13 presents the use and non-use value per hectare based on the number of day trips per year and the number of hectares open for visitors.

For outdoor recreation, the value that will be used is the net economic value per recreation day of €31.76. These benefits can be assessed for various polders by estimating the number of visitors (day trips) per hectare of land that is open to recreation. The results are also presented in Table 8.13. Note that the benefits of recreation described in Section 8.3.3 are part of the economic benefits presented here.

8.5 Summary of modelling results

The goal of this section was to make a spatial–economic analysis of the benefits and costs of converting various (parts of) polders of the Vecht area according to three scenarios. In the Agriculture, Nature and Recreation scenarios, each polder is given a setting that depends on the current land use and the options for changing land use.

The economic model describes the changes in costs and benefits of the scenarios for each polder relative to the Reference scenario (i.e. the present situation continued). Table 8.14 gives a concise overview of the results of the NPV aggregated per region. The NPV of the Agricultural scenario is calculated from the increased benefits of intensifying agriculture in various polders. The

Table 8.14 *The net present value per region under the three scenarios*

Region	Changes relative to Reference scenario (million €)		
	Agricultural scenario	Nature scenario	Recreation scenario
North	76.99	−109.02	698.73
Middle	35.00	−34.20	741.61
South	63.60	−83.16	542.29
Total	175.59	−226.38	1982.63

intensification of agriculture has a positive effect on the NPV, because the benefits per hectare of agricultural land increase. Under the Nature scenario, part of the agricultural land is converted into nature. Under the Recreation scenario, the amount of agricultural land converted is the same as under the Nature scenario. The difference is that under the Recreational scenario, nature areas are open for recreation and some polders are to be used for water recreation. Under the Nature scenario, the costs of converting agricultural land are the only financial indicators on which the calculation of the NPV is based. Therefore, the NPV is negative: there are costs, but no financial benefits. In the Recreation scenario, some polders are converted into recreational areas: both land and water recreation. The NPV of the Recreation scenario is based on the costs of converting the land (including maintenance and opportunity costs) and the benefits of spending on recreation. The resulting NPV under the Recreation scenario is positive: the increase in recreational benefits (i.e. the spending on recreation) outweighs the costs of converting the land use. In the Nature and Recreation scenario, three phosphate-removal plants are installed to extract phosphates from the water in order to improve its quality and, therefore, indirectly the quality of nature.

Table 8.15 presents two of the environmental indicators for each scenario: the nitrogen and phosphorus surplus. Under the Agriculture scenario, the values of these indicators increase because in some polders agricultural land use is intensified (i.e. more nutrients per hectare are used). Under the Nature and Recreation scenarios, less nutrients are used and the nitrogen and phosphorus surpluses decrease. The amount of agricultural land converted under both scenarios is equal and, therefore, the reduction in the surpluses is equal.

Tables 8.14 and 8.15 show that intensifying of agriculture can be considered as positive in economic terms, but negative in environmental terms because of the increased surplus and run-off of nutrients. The Nature scenario is economically not very attractive, because of a reduction in NPV; however, the conversion from agriculture to nature turns out to be very positive for the environment. The Recreation scenario is the most attractive on both economic and nutrient

Table 8.15 *The surplus of nitrogen and phosphorus per region under the three scenarios*

Scenario	Changes in nutrient indicator relative to Reference scenario (thousand kg)			
	Agriculture (surplus nitrogen)	Agriculture (surplus phosphate)	Nature/ Recreation (surplus nitrogen)	Nature/ Recreation (surplus phosphate)[a]
North	1260.3	209.5	−494.05	−90.20
Middle	573.8	95.2	−156.65	−28.60
South	1041.1	173.1	−352.46	−64.35
Total	2875.2	477.8	−1003.16	−183.15

[a] Under the Nature and Recreation scenarios, the nutrient indicators have the same values.

terms, because it has a positive NPV and the nutrient flow is reduced. Note that these conclusions do not take into account the benefits of nature conservation or the feedback to economic benefits of improved environmental quality. The final evaluation will combine the economic indicators with environmental quality and spatial equity indicators. Only then will a complete evaluation and ranking of the scenarios will be possible. This section has shown that the spatial–economic model of the Vecht area provides useful information about the impacts of converting land use.

Appendix 8.1 Calculation of economic and nutrient variables under the Agriculture scenario

The economic data, total revenue and total employment per hectare and nutrient data are calculated using the following equations, which essentially reflect a straightforward accounting procedure. Figures 8.3 and 8.4 present these calculations graphically.

Total revenue per hectare

Milk production per dairy cow (kg) × Milk price received (€/kilogram) = Milk revenues per dairy cow (€)

Other revenues per dairy cow (€) + Milk revenues per dairy cow (€) = Total revenues per dairy cow (€)

Other revenues per farm (€)/Average farm size (ha) = Other revenues per ha (€)

Total revenues per dairy cow (€) × Number of dairy cows per ha + Other revenues per ha (€) = Total revenues per ha (€)

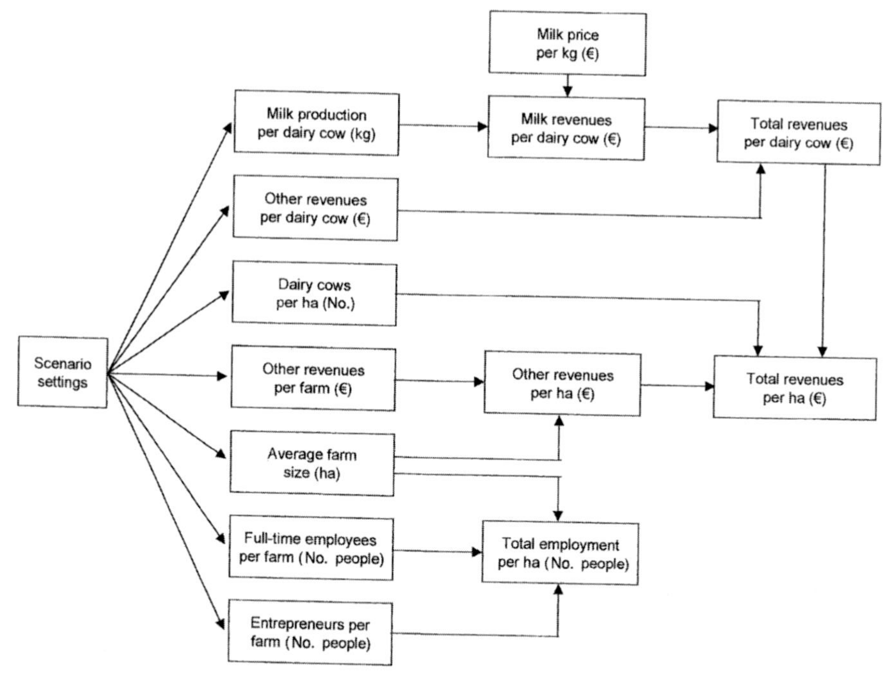

Fig. 8.3 Calculation of economic variables for agriculture.

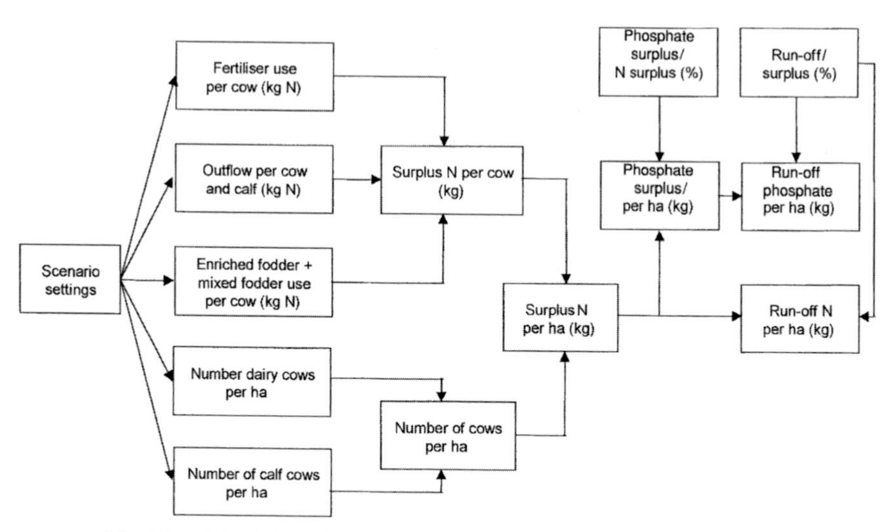

Fig. 8.4 Calculation of nutrient variables. N, nitrogen.

Total employment per hectare

(Full-time employees per farm + Full-time entrepreneurs per farm)/
Average farm size (ha) = Total employment per ha

The surplus of nitrogen per hectare

Fertiliser use per cow (kg N) + Enriched and mixed fodder user per cow (kg N) – Outflow per cow (kg N) = Nitrogen surplus per cow

Calf cows + Dairy cows = Cows per ha

Nitrogen surplus per cow (kg) × Cows per ha = Nitrogen surplus per ha (kg)

The surplus of phosphate per hectare

Nitrogen surplus per ha (kg) × Percentage (phosphate surplus/nitrogen surplus) = Phosphate surplus per ha (kg)

The run-off of nitrogen per hectare

Surplus nitrogen per ha (kg) × Run-off percentage = Run-off of nitrogen per ha (kg)

The run-off of phosphate per hectare

Surplus phosphate per ha (kg) × Run-off percentage = Run-off of phosphate per ha (kg)

Appendix 8.2 Outdoor recreation data

Tables 8.16 and 8.17 give estimates for short and long holidays in the Vecht area.

Table 8.16 *Estimated numbers of short holidays*

Variable	Quantity	Unit	Year of data	Source
Number of short holidays in 'Utrecht-Holland lake area'	230 000	Times	1995	CBS (1996b)
Part of area of 'Utrecht-Holland lake area' that is part of Vecht area	0.215	Part	1996	13 375/62 265 km^2 CBS (1996c)
Number of short holidays in 'Utrechtse Heuvelrug' and 'Het Gooi area'	230 000	Times	1995	CBS (1996b)
Part of area of Utrecht-Holland lake area that is part of Vecht area	0.442	Part	1996	29 282/66 255 km^2 CBS (1996c)
Area of 'rest of Vecht area'	9349[a]	km^2	1995/1996	52 006 − 13 375 − 29 282
Number of short holidays in Vecht area	184 164	Times		151 057 × 52 006/(13 375 + 29 282)
Number of short holidays per hectare	3.541	Times	1995/1996	184 164/52 006
Number of short holidays per hectare of forest/recreation and open water	30.191	Times	1995/1996	184 164/6100[b]

[a]The total area of the Vecht area is 52 006 km^2 (52 006 ha) (CBS, 1996c).
[b]Based on the number of short holidays per hectare for recreation in the Reference scenario (1200 ha forest/nature and 4900 ha open water; 6100 ha in total).

Table 8.17 *Estimated numbers of long holidays*

Variable	Quantity	Source
Number of long holidays in 'Utrecht-Holland lake area'	140 000	CBS (1996b)
Number of long holidays in 'Utrechtse Heuvelrug' and 'Het Gooi area'	230 000	CBS (1996b)
Number of long holidays in Vecht area	127 045	104 206 × 52 006/(13 375 + 29 282)
Number of long holidays in Vecht area per hectare	2.443	127 045/52 006
Number of long holidays in Vecht area per hectare for recreation	20.83	127 045/6100

Appendix 8.3 Economic scenario results on a polder level

Tables 8.18–8.20 give the economic and environmental outputs for each polder under the Agriculture, Nature and Recreation scenarios, respectively, relative to the Reference scenario.

Table 8.18 *The economic and environmental output under the Agriculture scenario (changes relative to Reference scenario)*

Polder	Setting	Agricultural land (ha) (base)	Agricultural land (ha) (change)	Total revenue (€)	Employment (No. people)	Dairy cows	Total net present value (5% interest, 10 year; €)[a]	Surplus nitrogen (kg)	Surplus phosphate (kg)	Run-off nitrogen (kg)	Run-off phosphate (kg)
1	2	350	350	1 103 503	−3.08	329	8 520 956	139 475	23 188	20 031	3 570
2	2	250	250	788 216	−2.20	235	6 086 397	99 625	16 563	14 308	2 550
3	2	200	200	630 573	−1.76	188	4 869 117	79 700	13 250	11 446	2 040
4	0	75	0	0	0.00	0	0	0	0	0	0
5	2	250	250	788 216	−2.20	235	6 086 397	99 625	16 563	14 308	2 550
6	0	100	0	0	0.00	0	0	0	0	0	0
7	2	150	150	472 930	−1.32	141	3 651 838	59 775	9 938	8 585	1 530
8	2	325	325	1 024 681	−2.86	306	7 912 316	129 513	21 531	18 600	3 315
9	2	125	125	394 108	−1.10	118	3 043 198	49 813	8 281	7 154	1 275
10	2	125	125	394 108	−1.10	118	3 043 198	49 813	8 281	7 154	1 275
11	0	175	0	0	0.00	0	0	0	0	0	0
12	0	75	0	0	0.00	0	0	0	0	0	0
13	2	25	25	78 822	−0.22	24	608 640	9 963	1 656	1 431	255
14	1	150	75	236 465	−0.66	71	1 825 919	29 888	4 969	4 292	765
15	2	125	125	394 108	−1.10	118	3 043 198	49 813	8 281	7 154	1 275
16	0	0	0	0	0.00	0	0	0	0	0	0
17	2	0	0	0	0.00	0	0	0	0	0	0
18	1	200	100	315 286	−0.88	94	2 434 559	39 850	6 625	5 723	1 020
19	2	50	50	157 643	−0.44	47	1 217 279	19 925	3 313	2 862	510
20	2	425	425	1 339 968	−3.74	400	10 346 875	169 363	28 156	24 323	4 335
21	0	375	0	0	0.00	0	0	0	0	0	0
22	0	25	0	0	0.00	0	0	0	0	0	0
23	1	175	87.5	275 876	−0.77	82	2 130 239	34 869	5 797	5 008	893

Table 8.18 (cont.)

Polder	Setting	Agricultural land (ha) (base)	Agricultural land (ha) (change)	Total revenue (€)	Employment (No. people)	Dairy cows	Total net present value (5% interest, 10 year; €)[a]	Surplus nitrogen (kg)	Surplus phosphate (kg)	Run-off nitrogen (kg)	Run-off phosphate (kg)
24	2	125	125	394 108	−1.10	118	3 043 198	49 813	8 281	7 154	1 275
25	2	75	75	236 465	−0.66	71	1 825 919	29 888	4 969	4 292	765
26	2	150	150	472 930	−1.32	141	3 651 838	59 775	9 938	8 585	1 530
27	2	200	200	630 573	−1.76	188	4 869 117	79 700	13 250	11 446	2 040
28	1	200	100	315 286	−0.88	94	2 434 559	39 850	6 625	5 723	1 020
29	0	0	0	0	0.00	0	0	0	0	0	0
30	0	100	0	0	0.00	0	0	0	0	0	0
31	2	75	75	236 465	−0.66	71	1 825 919	29 888	4 969	4 292	765
32	1	50	25	78 822	−0.22	24	608 640	9 963	1 656	1 431	255
33	0	0	0	0	0.00	0	0	0	0	0	0
34	0	0	0	0	0.00	0	0	0	0	0	0
35	0	0	0	0	0.00	0	0	0	0	0	0
36	2	150	150	472 930	−1.32	141	3 651 838	59 775	9 938	8 585	1 530
37	2	125	125	394 108	−1.10	118	3 043 198	49 813	8 281	7 154	1 275
38	2	275	275	867 038	−2.42	259	6 695 037	109 588	18 219	15 738	2 805
39	1	225	112.5	354 697	−0.99	106	2 738 879	44 831	7 453	6 438	1 148
40	1	200	100	315 286	−0.88	94	2 434 559	39 850	6 625	5 723	1 020
41	1	325	162.5	512 341	−1.43	153	3 956 158	64 756	10 766	9 300	1 658
42	2	275	275	867 038	−2.42	259	6 695 037	109 588	18 219	15 738	2 805
43	2	350	350	1 103 503	−3.08	329	8 520 956	139 475	23 188	20 031	3 570
44	2	200	200	630 573	−1.76	188	4 869 117	79 700	13 250	11 446	2 040
45	2	100	100	315 286	−0.88	94	2 434 559	39 850	6 625	5 723	1 020
46	2	275	275	867 038	−2.42	259	6 695 037	109 588	18 219	15 738	2 805
47	2	175	175	551 751	−1.54	165	4 260 478	69 738	11 594	10 015	1 785

48	0	0	0	0	0.00	0	0	0	0	0	0
49	2	25	25	78 822	−0.22	24	608 640	9 963	1 656	1 431	255
50	0	0	0	0	0.00	0	0	0	0	0	0
51	0	50	0	0	0.00	0	0	0	0	0	0
52	1	300	150	472 930	−1.32	141	3 651 838	59 775	9 938	8 585	1 530
53	2	100	100	315 286	−0.88	94	2 434 559	39 850	6 625	5 723	1 020
54	2	75	75	236 465	−0.66	71	1 825 919	29 888	4 969	4 292	765
55	2	25	25	78 822	−0.22	24	608 640	9 963	1 656	1 431	255
56	0	0	0	0	0.00	0	0	0	0	0	0
57	2	225	225	709 395	−1.98	212	5 477 757	89 663	14 906	12 877	2 295
58	2	25	25	78 822	−0.22	24	608 640	9 963	1 656	1 431	255
59	0	500	0	0	0.00	0	0	0	0	0	0
60	2	75	75	236 465	−0.66	71	1 825 919	29 888	4 969	4 292	765
61	0	0	0	0	0.00	0	0	0	0	0	0
62	2	0	0	0	0.00	0	0	0	0	0	0
63	2	150	150	472 930	−1.32	141	3 651 838	59 775	9 938	8 585	1 530
64	2	100	100	315 286	−0.88	94	2 434 559	39 850	6 625	5 723	1 020
65	0	0	0	0	0.00	0	0	0	0	0	0
66	2	75	75	236 465	−0.66	71	1 825 919	29 888	4 969	4 292	765
67	2	150	150	472 930	−1.32	141	3 651 838	59 775	9 938	8 585	1 530
68	2	100	100	315 286	−0.88	94	2 434 559	39 850	6 625	5 723	1 020
69	2	150	150	472 930	−1.32	141	3 651 838	59 775	9 938	8 585	1 530
70	2	0	0	0	0.00	0	0	0	0	0	0
71	0	0	0	0	0.00	0	0	0	0	0	0
72	1	150	75	236 465	−0.66	71	1 825 919	29 888	4 969	4 292	765
73	0	150	0	0	0.00	0	0	0	0	0	0
Total[b]		9 825	7 400	22 740 038	−63.47	6 780	175 592 549	2 874 181	477 828	412 771	73 568

[a] Total net present value equals the net present value of the benefits, because there are no costs in this scenario.

[b] Variation between exact summation in columns and Totals result from the rounding-off processes used.

Table 8.19 *The economic and environmental output under the Nature scenario (changes relative to Reference scenario)*

Polder	Setting	Agricultural land (ha) (base)	Agricultural land (ha) (change)	Acquiring costs (€)	Restructuring costs (€)	Maintenance costs (€)	Opportunity costs (€)	Employment (No. people)	Phosphate-removal plants		Total net present value (5%, 10 year; €)[a]	Surplus nitrogen (kg)	Surplus phosphate (kg)	Run-off nitrogen (kg)	Run-off phosphate (kg)
									Fixed costs (€)	Variable costs (€)					
1	0	350	0	0	0	0	0	0	0	0	0	0	0	0	0
2	0	250	0	0	0	0	0	0	0	0	0	0	0	0	0
3	0	200	0	0	0	0	0	0	0	0	0	0	0	0	0
4	0	75	0	0	0	0	0	0	0	0	0	0	0	0	0
5	1	250	125	2 608 272	1 272 570	58 793	294 276	−13	0	0	−6 422 348	−30 125	−5 500	−4 641	−848
6	0	100	0	0	0	0	0	0	32 200	60 000	−493 971	0	0	0	0
7	2	150	150	3 129 926	1 527 084	70 551	353 132	−15	0	0	−7 706 818	−36 150	−6 600	−5 570	−1 017
8	1	325	162.5	3 390 754	1 654 341	76 431	382 559	−16	0	0	−8 349 052	−39 163	−7 150	−6 034	−1 102
9	2	125	125	2 608 272	1 272 570	58 793	294 276	−13	0	0	−6 422 348	−30 125	−5 500	−4 641	−848
10	2	125	125	2 608 272	1 272 570	58 793	294 276	−13	0	0	−6 422 348	−30 125	−5 500	−4 641	−848
11	0	175	0	0	0	0	0	0	0	0	0	0	0	0	0
12	0	75	0	0	0	0	0	0	0	0	0	0	0	0	0
13	1	25	12.5	260 827	127 257	5 879	29 428	−1	0	0	−642 235	−3 013	−550	−464	−85
14	2	150	150	3 129 926	1 527 084	70 551	353 132	−15	0	0	−7 706 818	−36 150	−6 600	−5 570	−1 017
15	0	125	0	0	0	0	0	0	0	0	0	0	0	0	0
16	0	0	0	0	0	0	0	0	0	0	0	0	0	0	0
17	1	0	0	0	0	0	0	0	0	0	0	0	0	0	0
18	2	200	200	4 173 235	2 036 112	94 069	470 842	−20	0	0	−10 275 757	−48 200	−8 800	−7 426	−1 356
19	0	50	0	0	0	0	0	0	0	0	0	0	0	0	0
20	2	425	425	8 868 125	4 326 738	199 896	1 000 540	−43	0	0	−21 835 983	−102 425	−18 700	−15 780	−2 882
21	1	375	187.5	3 912 408	1 908 855	88 189	441 415	−19	54 500	97 100	−10 435 207	−45 188	−8 250	−6 962	−1 271
22	0	25	0	0	0	0	0	0	0	0	0	0	0	0	0
23	1	175	87.5	1 825 790	890 799	41 155	205 994	−9	54 500	97 100	−5 297 329	−21 088	−3 850	−3 249	−593
24	1	125	62.5	1 304 136	636 285	29 396	147 138	−6	0	0	−3 211 174	−15 063	−2 750	−2 321	−424
25	1	75	37.5	782 482	381 771	17 638	88 283	−4	54 500	97 100	−2 728 390	−9 038	−1 650	−1 392	−254
26	0	150	0	0	0	0	0	0	54 500	97 100	−801 685	0	0	0	0
27	1	200	100	2 086 618	1 018 056	47 034	235 421	−10	54 500	97 100	−5 939 564	−24 100	−4 400	−3 713	−678
28	1	200	100	2 086 618	1 018 056	47 034	235 421	−10	0	0	−5 137 878	−24 100	−4 400	−3 713	−678
29	1	0	0	0	0	0	0	0	0	0	0	0	0	0	0
30	2	100	100	2 086 618	1 018 056	47 034	235 421	−10	0	0	−5 137 878	−24 100	−4 400	−3 713	−678
31	1	75	37.5	782 482	381 771	17 638	88 283	−4	0	0	−1 926 704	−9 038	−1 650	−1 392	−254
32	1	50	25	521 654	254 514	11 759	58 855	−3	0	0	−1 284 470	−6 025	−1 100	−928	−170
33	0	0	0	0	0	0	0	0	0	0	0	0	0	0	0
34	0	0	0	0	0	0	0	0	0	0	0	0	0	0	0
35	1	0	0	0	0	0	0	0	0	0	0	0	0	0	0

36	0	150	0	0	0	0	0	0	0	0	0	0	0	0	0	
37	0	125	0	0	0	0	0	0	0	0	0	0	0	0	0	
38	0	275	0	0	0	0	0	0	0	0	0	0	0	0	0	
39	2	225	225	4 694 890	2 290 626	105 827	529 698	−23	32 059	57 118	−12 031 806	−54 225	−9 900	−8 354	−1 526	
40	2	200	200	4 173 235	2 036 112	94 069	470 842	−20	32 059	57 118	−10 747 336	−48 200	−8 800	−7 426	−1 356	
41	1	325	162.5	3 390 754	1 654 341	76 431	382 559	−16	0	0	−8 349 052	−39 163	−7 150	−6 034	−1 102	
42	1	275	137.5	2 869 099	1 399 827	64 672	323 704	−14	32 059	57 118	−7 536 162	−33 138	−6 050	−5 105	−932	
43	0	350	0	0	0	0	0	0	32 059	57 118	−471 580	0	0	0	0	
44	0	200	0	0	0	0	0	0	32 059	57 118	−471 580	0	0	0	0	
45	0	100	0	0	0	0	0	0	32 059	57 118	−471 580	0	0	0	0	
46	1	275	137.5	2 869 099	1 399 827	64 672	323 704	−14	0	0	−7 064 583	−33 138	−6 050	−5 105	−932	
47	2	175	175	3 651 581	1 781 598	82 310	411 987	−18	0	0	−8 991 287	−42 175	−7 700	−6 498	−1 187	
48	0	0	0	0	0	0	0	0	0	0	0	0	0	0	0	
49	1	25	12.5	260 827	127 257	5 879	29 428	−1	0	0	−642 235	−3 013	−550	−464	−85	
50	0	0	0	0	0	0	0	0	32 059	57 118	−471 580	0	0	0	0	
51	0	50	0	0	0	0	0	0	0	0	0	0	0	0	0	
52	2	300	300	6 259 853	3 054 168	141 103	706 264	−30	0	0	−15 413 635	−72 300	−13 200	−11 139	−2 034	
53	1	100	50	1 043 309	509 028	23 517	117 711	−5	32 059	57 118	−3 040 519	−12 050	−2 200	−1 857	−339	
54	0	75	0	0	0	0	0	0	0	0	0	0	0	0	0	
55	0	25	0	0	0	0	0	0	0	0	0	0	0	0	0	
56	1	0	0	0	0	0	0	0	0	0	0	0	0	0	0	
57	2	225	225	4 694 890	2 290 626	105 827	529 698	−23	32 059	57 118	−12 031 806	−54 225	−9 900	−8 354	−1 526	
58	1	25	12.5	260 827	127 257	5 879	29 428	−1	32 059	57 118	−1 113 814	−3 013	−550	−464	−85	
59	0	500	0	0	0	0	0	0	0	0	0	0	0	0	0	
60	1	75	37.5	782 482	381 771	17 638	88 283	−4	32 059	57 118	−2 398 284	−9 038	−1 650	−1 392	−254	
61	0	0	0	0	0	0	0	0	0	0	0	0	0	0	0	
62	1	0	0	0	0	0	0	0	32 059	57 118	−471 580	0	0	0	0	
63	0	150	0	0	0	0	0	0	32 059	57 118	−471 580	0	0	0	0	
64	0	100	0	0	0	0	0	0	32 059	57 118	−471 580	0	0	0	0	
65	0	0	0	0	0	0	0	0	32 059	57 118	−471 580	0	0	0	0	
66	0	75	0	0	0	0	0	0	0	0	0	0	0	0	0	
67	0	150	0	0	0	0	0	0	0	0	0	0	0	0	0	
68	1	100	50	1 043 309	509 028	23 517	117 711	−5	32 059	57 118	−3 040 519	−12 050	−2 200	−1 857	−339	
69	1	150	75	1 564 963	763 542	35 276	176 566	−8	32 059	57 118	−4 324 988	−18 075	−3 300	−2 785	−509	
70	0	0	0	0	0	0	0	0	0	0	0	0	0	0	0	
71	0	0	0	0	0	0	0	0	0	0	0	0	0	0	0	
72	2	150	150	3 129 926	1 527 084	70 551	353 132	−15	0	0	−7 706 818	−36 150	−6 600	−5 570	−1 017	
73	0	150	0	0	0	0	0	0	0	0	0	0	0	0	0	
Total		9 825	4 162.5	86 855 456	42 376 577	1 957 804	9 799 406	−417	849 700	1 516 500	−226 383 436	−1 003 163	−183 150	−154 554	−28 222	

[a] Total net present value equals the net present value of the costs, because in this scenario there are no benefits.

[b] Variation between exact summation in columns and Totals result from the rounding-off processes used.

Table 8.20 *The economic and environmental output under the Recreation scenario*

Polder	Setting	Agricultural land (ha) (base)	Agricultural land (ha) (change)	Open water (ha)	Acquiring costs (€)	Restructuring costs (€)	Maintenance costs (€)	Opportunity costs (€)
1	0	350	0	0	0	0	0	0
2	0	250	0	0	0	0	0	0
3	0	200	0	0	0	0	0	0
4	0	75	0	0	0	0	0	0
5	1	250	125	0	2 608 272	1 985 288	64 664	294 276
6	0	100	0	0	0	0	0	0
7	2	150	150	0	3 129 926	2 382 346	77 596	353 132
8	1	325	162.5	0	3 390 754	2 580 875	84 063	382 559
9	3	125	125	0	2 608 272	3 500 000	32 332	294 276
10	2	125	125	0	2 608 272	1 985 288	64 664	294 276
11	0	175	0	0	0	0	0	0
12	4	75	0	275	0	1 247 896	71 130	0
13	1	25	12.5	0	260 827	198 529	6 466	29 428
14	2	150	150	0	3 129 926	2 382 346	77 596	353 132
15	0	125	0	0	0	0	0	0
16	0	0	0	0	0	0	0	0
17	1	0	0	0	0	0	0	0
18	2	200	200	0	4 173 235	3 176 462	103 462	470 842
19	0	50	0	0	0	0	0	0
20	3	425	425	0	8 868 125	750 000	109 928	1 000 540
21	1	375	187.5	0	3 912 408	2 977 933	96 996	441 415
22	0	25	0	0	0	0	0	0
23	1	175	87.5	0	1 825 790	1 389 702	45 265	205 994
24	1	125	62.5	0	1 304 136	992 644	32 332	147 138
25	1	75	37.5	0	782 482	595 587	19 399	88 283
26	0	150	0	0	0	0	0	0
27	1	200	100	0	2 086 618	1 588 231	51 731	235 421
28	1	200	100	0	2 086 618	1 588 231	51 731	235 421
29	4	0	0	375	0	1 701 676	96 996	0
30	2	100	100	0	2 086 618	1 588 231	51 731	235 421
31	1	75	37.5	0	782 482	595 587	19 399	88 283
32	1	50	25	0	521 654	397 058	12 933	58 855
33	0	0	0	0	0	0	0	0
34	0	0	0	0	0	0	0	0
35	4	0	0	1 225	0	5 558 808	316 852	0
36	0	150	0	0	0	0	0	0
37	0	125	0	0	0	0	0	0
38	0	275	0	0	0	0	0	0
39	2	225	225	0	4 694 890	3 573 519	116 395	529 698
40	2	200	200	0	4 173 235	3 176 462	103 462	470 842
41	1	325	162.5	0	3 390 754	2 580 875	84 063	382 559
42	1	275	137.5	0	2 869 099	2 183 817	71 130	323 704
43	0	350	0	0	0	0	0	0
44	0	200	0	0	0	0	0	0
45	0	100	0	0	0	0	0	0
46	1	275	137.5	0	2 869 099	2 183 817	71 130	323 704
47	2	175	175	0	3 651 581	2 779 404	90 529	411 987
48	0	0	0	0	0	0	0	0
49	1	25	12.5	0	260 827	198 529	6 466	29 428
50	0	0	0	0	0	0	0	0

(changes relative to Reference scenario)

Benefits (€)	Employment (No. people)	Net present value (5% interest, 10 years; €)			Surplus nitrogen (kg)	Surplus phosphate (kg)	Run-off nitrogen (kg)	Run-off phosphate (kg)
		Total	Costs	Benefits				
0	0	0	0	0	0	0	0	0
0	0	0	0	0	0	0	0	0
0	0	0	0	0	0	0	0	0
0	0	0	0	0	0	0	0	0
5 681 555	112	36 725 003	−7 146 460	43 871 463	−30 125	−5 500	−4 641	−848
0	0	−493 971	−493 971	0	0	0	0	0
6 817 866	135	44 070 004	−8 575 752	52 645 756	−36 150	−6 600	−5 570	−1 017
7 386 022	146	47 742 504	−9 290 398	57 032 902	−39 163	−7 150	−6 034	−1 102
5 681 555	112	35 532 079	−8 339 385	43 871 463	−30 125	−5 500	−4 641	−848
5 681 555	112	36 725 003	−7 146 460	43 871 463	−30 125	−5 500	−4 641	−848
0	0	0	0	0	0	0	0	0
12 499 421	0	94 779 500	−1 737 719	96 517 219	0	0	0	0
568 156	11	3 672 500	−714 646	4 387 146	−3 013	−550	−464	−85
6 817 866	135	44 070 004	−8 575 752	52 645 756	−36 150	−6 600	−5 570	−1 017
0	0	0	0	0	0	0	0	0
0	0	0	0	0	0	0	0	0
0	0	0	0	0	0	0	0	0
9 090 488	180	58 760 005	−11 434 336	70 194 341	−48 200	−8 800	−7 426	−1 356
0	0	0	0	0	0	0	0	0
19 317 288	382	131 428 115	−17 734 860	149 162 975	−102 425	−18 700	−15 780	−2 882
8 522 333	169	54 285 819	−11 521 375	65 807 195	−45 188	−8 250	−6 962	−1 271
0	0	0	0	0	0	0	0	0
3 977 089	79	24 905 817	−5 804 207	30 710 024	−21 088	−3 850	−3 249	−593
2 840 778	56	18 362 502	−3 573 230	21 935 732	−15 063	−2 750	−2 321	−424
1 704 467	34	10 215 816	−2 945 623	13 161 439	−9 038	−1 650	−1 392	−254
0	0	−801 685	−801 685	0	0	0	0	0
4 545 244	90	28 578 317	−6 518 853	35 097 171	−24 100	−4 400	−3 713	−678
4 545 244	90	29 380 002	−5 717 168	35 097 171	−24 100	−4 400	−3 713	−678
17 044 666	0	129 244 772	−2 369 617	131 614 390	0	0	0	0
4 545 244	90	29 380 002	−5 717 168	35 097 171	−24 100	−4 400	−3 713	−678
1 704 467	34	11 017 501	−2 143 938	13 161 439	−9 038	−1 650	−1 392	−254
1 136 311	22	7 345 001	−1 429 292	8 774 293	−6 025	−1 100	−928	−170
0	0	0	0	0	0	0	0	0
0	0	0	0	0	0	0	0	0
55 679 241	0	422 199 589	−7 740 750	429 940 339	0	0	0	0
0	0	0	0	0	0	0	0	0
0	0	0	0	0	0	0	0	0
0	0	0	0	0	0	0	0	0
10 226 799	202	65 633 426	−13 335 208	78 968 634	−54 225	−9 900	−8 354	−1 526
9 090 488	180	58 288 425	−11 905 916	70 194 341	−48 200	−8 800	−7 426	−1 356
7 386 022	146	47 742 504	−9 290 398	57 032 902	−39 163	−7 150	−6 034	−1 102
6 249 711	124	39 925 924	−8 332 686	48 258 610	−33 138	−6 050	−5 105	−932
0	0	−471 580	−471 580	0	0	0	0	0
0	0	−471 580	−471 580	0	0	0	0	0
0	0	−471 580	−471 580	0	0	0	0	0
6 249 711	124	40 397 503	−7 861 106	48 258 610	−33 138	−6 050	−5 105	−932
7 954 177	157	51 415 004	−10 005 044	61 420 048	−42 175	−7 700	−6 498	−1 187
0	0	0	0	0	0	0	0	0
568 156	11	3 672 500	−714 646	4 387 146	−3 013	−550	−464	−85
0	0	−471 580	−471 580	0	0	0	0	0

Table 8.20 (cont.)

Polder	Setting	Agricultural land (ha) (base)	Agricultural land (ha) (change)	Open water (ha)	Acquiring costs (€)	Restructuring costs (€)	Maintenance costs (€)	Opportunity costs (€)
51	0	50	0	0	0	0	0	0
52	2	300	300	0	6 259 853	4 764 692	155 193	706 264
53	1	100	50	0	1 043 309	794 115	25 865	117 711
54	0	75	0	0	0	0	0	0
55	0	25	0	0	0	0	0	0
56	4	0	0	350	0	1 588 231	90 529	0
57	2	225	225	0	4 694 890	3 573 519	116 395	529 698
58	1	25	12.5	0	260 827	198 529	6 466	29 428
59	0	500	0	0	0	0	0	0
60	1	75	37.5	0	782 482	595 587	19 399	88 283
61	0	0	0	0	0	0	0	0
62	1	0	0	0	0	0	0	0
63	0	150	0	0	0	0	0	0
64	0	100	0	0	0	0	0	0
65	0	0	0	0	0	0	0	0
66	0	75	0	0	0	0	0	0
67	0	150	0	0	0	0	0	0
68	1	100	50	0	1 043 309	794 115	25 865	117 711
69	1	150	75	0	1 564 963	1 191 173	38 798	176 566
70	0	0	0	0	0	0	0	0
71	0	0	0	0	0	0	0	0
72	2	150	150	0	3 129 926	2 382 346	77 596	353 132
73	0	150	0	0	0	0	0	0
Total		9 825	4 162.5	2 225	86 855 456	71 721 446	2 586 547	9 799 406

[a] Variation between exact summation in columns and Totals result from the rounding-off processes used.

Benefits (€)	Employment (No. people)	Net present value (5% interest, 10 years; €)			Surplus nitrogen (kg)	Surplus phosphate (kg)	Run-off nitrogen (kg)	Run-off phosphate (kg)
		Total	Costs	Benefits				
0	0	0	0	0	0	0	0	0
13 635 732	270	88 140 007	−17 151 504	105 291 512	−72 300	−13 200	−11 139	−2 034
2 272 622	45	14 218 422	−3 330 164	17 548 585	−12 050	−2 200	−1 857	−339
0	0	0	0	0	0	0	0	0
0	0	0	0	0	0	0	0	0
15 908 355	0	120 628 454	−2 211 643	122 840 097	0	0	0	0
10 226 799	202	65 633 426	−13 335 208	78 968 634	−54 225	−9 900	−8 354	−1 526
568 156	11	3 200 921	−1 186 226	4 387 146	−3 013	−550	−464	−85
0	0	0	0	0	0	0	0	0
1 704 467	34	10 545 921	−2 615 518	13 161 439	−9 038	−1 650	−1 392	−254
0	0	0	0	0	0	0	0	0
0	0	−471 580	−471 580	0	0	0	0	0
0	0	−471 580	−471 580	0	0	0	0	0
0	0	−471 580	−471 580	0	0	0	0	0
0	0	−471 580	−471 580	0	0	0	0	0
0	0	0	0	0	0	0	0	0
0	0	0	0	0	0	0	0	0
2 272 622	45	14 218 422	−3 330 164	17 548 585	−12 050	−2 200	−1 857	−339
3 408 933	67	21 563 422	−4 759 456	26 322 878	−18 075	−3 300	−2 785	−509
0	0	0	0	0	0	0	0	0
0	0	0	0	0	0	0	0	0
6 817 866	135	44 070 004	−8 575 752	52 645 756	−36 150	−6 600	−5 570	−1 017
0	0	0	0	0	0	0	0	0
290 327 470	3 745	1 982 645 848	−259 185 921	2 241 831 770	−1 003 163	−183 150	−154 554	−28 222

9

Performance indicators for the evaluation

9.1 Introduction

The evaluation framework (Fig. 9.1) has the task of ranking scenarios according to their performance on three objectives.

1. Net present value (NPV) as a proxy for economic efficiency, the conventional objective in project evaluation.
2. Spatial equity, defined as the distribution across space of gains and losses, both economic and ecological, from management choices.
3. Environmental quality, reflecting ecological functioning and responding to the 'precautionary principle' (e.g. Pearce and Perrings, 1995), which highlights uncertainties regarding the complexity of hydrological and ecological processes.

It is argued that such a set of multiple objectives will make the trade-off among different policy objectives more transparent.

The information on which the evaluation will be based is provided by output from the ecological–hydrological and economic models. These models have been described in Chapters 7 and 8. The economic model generates NPV for each scenario and so performance indicators for this evaluation objective do not need development. The aim of this section is to develop performance indicators for environmental quality and spatial equity. We discuss performance indicators for environmental quality and for spatial equity. The evaluation itself is presented in the next chapter.

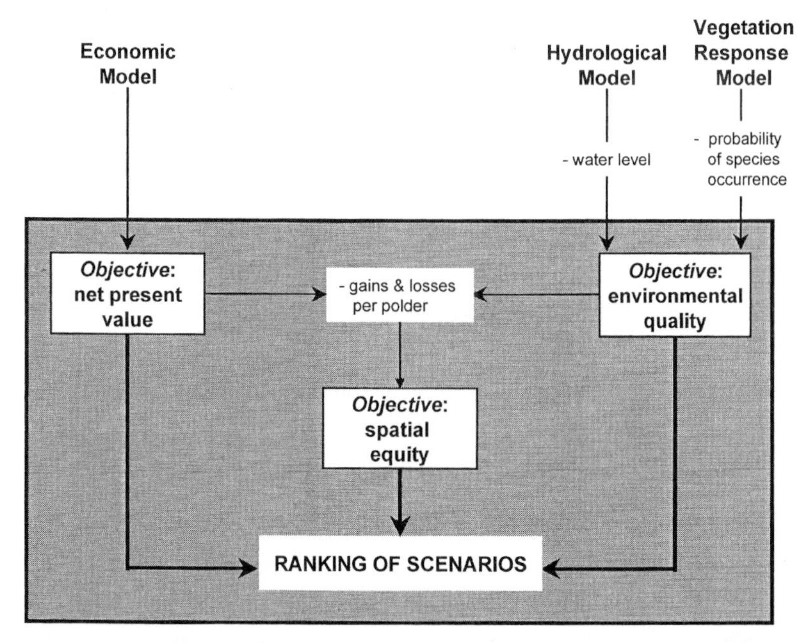

Fig. 9.1 The evaluation framework.

9.2 Performance indicators for environmental quality

9.2.1 *Conceptual approach*

In line with Section 2.2.7, environmental quality may be reflected by three ecosystem characteristics:[1]

- ecosystem process;
- ecosystem structure; and
- ecosystem resilience.

The aim of this section is to develop performance indicators for each of these. These indicators will be based on output from the hydrological and ecological models.

The capturing of these three ecosystem characteristics for the remaining (semi-) natural wetlands of the Vecht in performance indicators is constrained by a number of factors. These include the specifics of the study area, past and present impacts on its wetlands, the type of analysis being undertaken, and the type of models being used.

[1] These characteristics draw from the literature on ecosystem health (e.g. Costanza *et al.*, 1992; Schaeffer *et al.*, 1988; Steedman, 1994), ecological sustainability (e.g. IUCN, UNEP & WWF, 1980; van Ierland and de Man, 1993), biological diversity (e.g. Ghilarov, 1996) and ecological integrity (e.g. Karr, 1991). See also Section 2.2.7.

The region is a product of interactions between human (largely agricultural) and natural systems over centuries. Wetlands have been lost, fragmented and put under stress by these influences. Human influence has focused on:

- the harvesting and thus physical removal of peat, usually followed by inundation;
- lowering of water levels to permit agricultural activities but also accelerating peat loss through its exposure to the atmosphere, with subsequent oxidation and decomposition;
- the abstraction of groundwater, causing an increased influence of Vecht river water on the wetlands of the region; and
- nutrient enrichment as a result of agricultural activities and of pollution of Vecht river water, with subsequent acceleration of the succession.

The wetland ecosystems of the Vecht are a product of gradients between fully aquatic and fully terrestrial conditions. Ecosystem development has led to a mosaic of lakes, marshes, reedbeds and forests with a high diversity in habitat and species. The rate of succession within this development has been accelerated and its direction is being diverted as a result of human activities and impacts. The typical 'climax' towards which succession would move the system consists of a sharp interface between aquatic and terrestrial ecosystems, both of which are less diverse (with respect to species, communities and landscapes) than the wetlands. Diversion of the successional series leads to communities comprising and even dominated by species atypical of these wetlands. Policies such as the *Vierde Nota over de Ruimtelijke Ordening (Extra)* (VINEX) (VROM, 1990), *Structuurschema Groene Ruimte* (LNV, 1996), *Natuurbeleidsplan* (LNV, 1990) and the *Derde Nota Waterhuishouding* (V&W, 1989), as well as actions being taken for spatial planning (*Landinrichtinsbeleid*) show that these changes are considered undesirable.

The vegetation response model used in this case study, although providing a detailed breakdown across plant species, is static. Consequently, there is no attempt to represent ecosystem processes and their dynamics. This provides major constraints on reflecting the three ecosystem characteristics, which are dynamic. More specifically:

- with regard to ecosystem processes, the model does not estimate biomass or primary productivity and does not capture crucial ecosystem processes behind the fixation and cycling of carbon, nutrients, etc.;
- with regard to ecosystem structure, the model indicates plant species diversity but does not reflect higher trophic levels or interactions among species; and

- with regard to resilience, the absence of dynamics means that the time taken for recovery from disturbance, the essential measure of resilience, cannot be estimated.

The output of the ecological and hydrological models comprises water levels, water quality and the likelihood of the occurrence of plant species based on these quantity and quality variables. This output, provided per 500 m $\times$ 500 m grid cell, must form the basis for the indicators being developed. The following discussion offers potential solutions (compromises) as to how these characteristics might be reflected.

An approach for representing the three ecosystem characteristics – process, structure and resilience – for this case study is offered in this section. This approach must address both the characteristics themselves and the boundaries discussed above.

Ecosystem process

Two ecosystem processes are of prime relevance to the functioning of the Vecht wetlands: primary production and peat accumulation. Primary production has been stimulated by nutrient enrichment, resulting in the invasion and even dominance of communities by plant species (e.g. *Calamagrostis canescens*, *Carex acutiformis* and *Carex riparia*) not normally so abundant under meso- and oligotrophic conditions. These species are termed 'eutrophic' species. Their probability of occurrence is likely to be stimulated by nutrient enrichment. Both ICHORS modules within the vegetation response model include such species, and one indicator for process can be based on these probabilities. Such an indicator makes a negative contribution to environmental quality and so will be assigned a negative value.

Peat accumulation not only reflects primary production but also its contribution to the ecosystem's essential structure and subsequent maintenance. Peat levels are a balance between peat formation and peat loss. The lowering of water levels, as occurs with polder management to promote agriculture, exposes peat to the atmosphere and triggers its oxidation and decomposition. The rate of peat loss with exposure may be assumed to exceed formation. Consequently, a threshold can be inferred for water levels below which peat decomposition exceeds formation and above which peat accumulation may occur.

Given sufficiently high water levels, peat formation and even accumulation may occur. This depends on a number of factors, although it is ultimately dependent on the presence of peat-forming species. Both water levels and the probability of the occurrence of peat-forming species are outputs from the models and are used to construct the second indicator for process. Such an indicator will

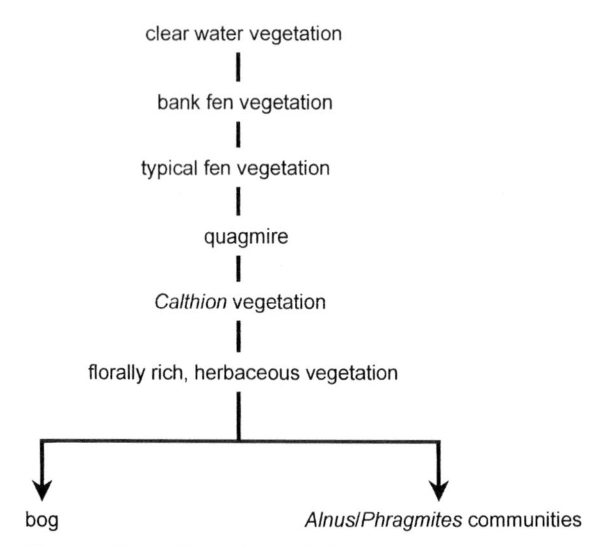

Fig. 9.2 Vegetation series typical of the succession in the Vecht wetlands.

make a positive contribution to environmental quality and so will be assigned a positive value.

Ecosystem structure

Ecosystem structure comprises not only species composition and diversity, but also trophic levels, functional groups and their multiple interactions. Since the vegetation response model focuses on the occurrence of plant species, plant diversity is the only aspect of structure that can be inferred. Biological diversity in the Vecht wetlands is largely a product of the species-rich successional stages separating fully aquatic and terrestrial ecosystems (see Fig. 9.2 for an approximation of the successional series). Fully aquatic and terrestrial ecosystems, in comparison with the intermediate wetland ecosystem, are much less rich in species. The succession can move towards two alternative climax communities, based on whether the wetlands develop a fen (mesotrophic) or bog (oligotrophic, acidic) character. Nutrient enrichment not only speeds up succession but also leads to alternative climax communities comprising atypical wetland species, namely acidic reedbeds instead of bogs, and reed-sedge vegetation instead of fens.

The main threat to diversity is accelerated succession in which this species-rich transitional zone becomes condensed into a smaller number of stages occupying a smaller space. Ideally, a measure of plant diversity would focus on an assessment per scenario of whether or not these successional stages were present and of the area that they covered. The spatial detail of the models could permit the latter, but a thorough search of the literature failed to yield reference statistics enabling the conversion of the model's output into the presence or absence of successional communities. However, plant species from these successional stages that are included in the vegetation response model can be identified. It is proposed, therefore, to assess plant diversity simply via the probability of the occurrence of these species per scenario. Such an indicator will make a positive contribution to environmental quality and so is assigned a positive value.

Resilience

There are two perspectives on resilience in the literature: the time taken for a system to return to its state prior to disturbance (Pimm, 1984) and the size of disturbance required to move a system out of one state and into another (Holling, 1973).

With regard to the former, two elements are required for the measurement of resilience: the state to which the system should return after disturbance and the time taken for this return. The state to which the system should return cannot be defined for the Vecht wetlands. They are in a state of perpetual but directed change as a result of succession. Physical intervention into the ecosystem, by human actions or by natural events (e.g. storms in which wave and wind action physically alter an area so that succession can begin again), is the only way in which a return to a previous state can occur. Further, the disturbance to the Vecht wetlands – brought about by nutrient enrichment – leads to the acceleration of this natural process. An alternative approach towards this perspective on resilience is to take the predisturbance rate of succession as the 'state' to which the system should return. The time taken to slow succession to this rate would then approximate recovery time and offer a measure of resilience. This is illustrated in Fig. 9.3a.

With regard to the second perspective on resilience, measurement tends to focus on the size of the disturbance. In the context of this study, disturbance has already effected a change and it is no longer possible to measure its size. However, an approach based on succession (as above) could focus on the redirection of a typical successional series towards a different climax. Measurement of this interpretation of resilience would focus on the presence of 'atypical' successional stages. This is shown in Fig. 9.3b.

The model available for this case study does not explicitly reflect succession and its underlying processes. Recovery cannot be simulated; recovery time cannot be estimated. Shifts between the various successional stages, as a result of the different measures undertaken in each scenario, can be indicated but cannot be accurately interpreted. Consequently, the interpretation in Fig. 9.3 could not be used.

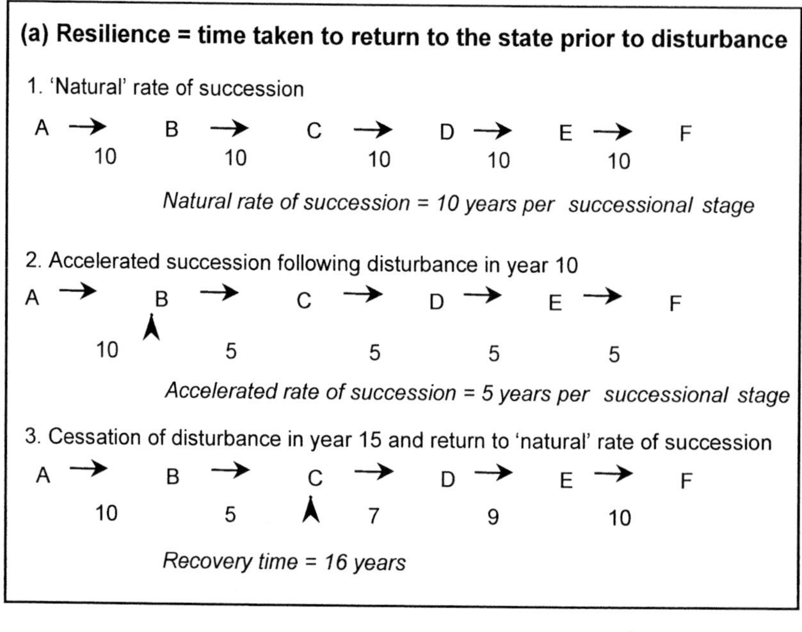

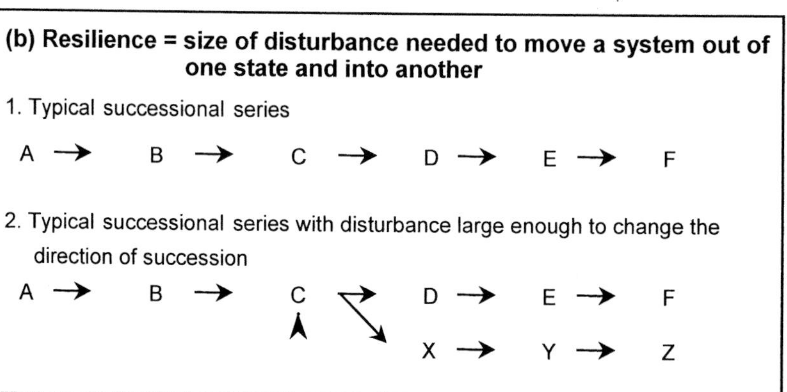

Fig. 9.3 Interpretation of 'resilience' for succession-driven ecosystems, as defined by (a) Pimm (1984) and (b) Holling (1973).

Nevertheless, the interpretation in Fig. 9.3 can be tentatively applied as follows. The successional series typical of the Vecht wetlands (see Fig. 9.2) has been disturbed by nutrient enrichment. Atypical species, generally favoured by nutrient-rich conditions, have become more prevalent in the wetlands and threaten to shift the direction of succession. Where surface waters determine abiotic conditions, a number of species indicative of a disturbed succession can be identified. Where groundwater still determines abiotic conditions, two alternative climax communities may develop: acidic reedbeds instead of bogs, and reed-sedge vegetation instead of *Alnus/Phragmites* communities. The acidic conditions necessary for bog (and, therefore, for acidic reedbed) development are relatively unlikely to occur, either naturally or as a result of the actions of local managers. Consequently, species from this alternative climax have been excluded from consideration. Plant species that are indicative of disturbed successional series can be identified from the two ecological modules. The indicator for resilience is based on the probabilities of occurrence of these species. Such an indicator will make a negative contribution to environmental quality and will be assigned a negative value. To avoid confusion, it will be referred to as the indicator of non-resilience.

The development of performance indicators and an aggregate index for environmental quality is summarised in Fig. 9.4 and elaborated on below. Development is focused on four aggregate measures representing the three ecosystem characteristics, using output from the hydrological and vegetation response models:

- ecosystem processes are reflected by an indicator for eutrophic plant species (*E*) and peat-forming species (*P*);
- ecosystem structure is reflected by an indicator for plant diversity (*B*) based on the presence of species belonging to vegetation types typical of these wetlands; and,
- non-resilience (*R*) is indicated by the extent to which the system has been pushed away from its normal successional sequence and rate.

9.2.2 *Construction of indicators and index*

This section presents the procedures for constructing these four performance indicators (*E*, *P*, *B* and *R*) and subsequently an index for environmental quality (EQ). The indicators are based primarily on output from the vegetation response model, which comprises the probability of the occurrence of a plant species per grid cell within the 73 polders of the Vecht region. The model comprises data sets for individual species, and these sets are grouped into two ecological modules, as discussed in Chapter 7. Some species appear in both ecological modules (with different data sets, because of the different water quality parameters being used); others appear two or more times in ICHORS2.0 with their data sets adjusted to reflect their capacity to dominate the vegetation. Examples of the latter include *Sphagnum* spp. and *Phragmites australis*. Indicator development means aggregation of this large volume of output. The approach taken for developing the indicators, and subsequently an index for environmental quality, is based on five steps.

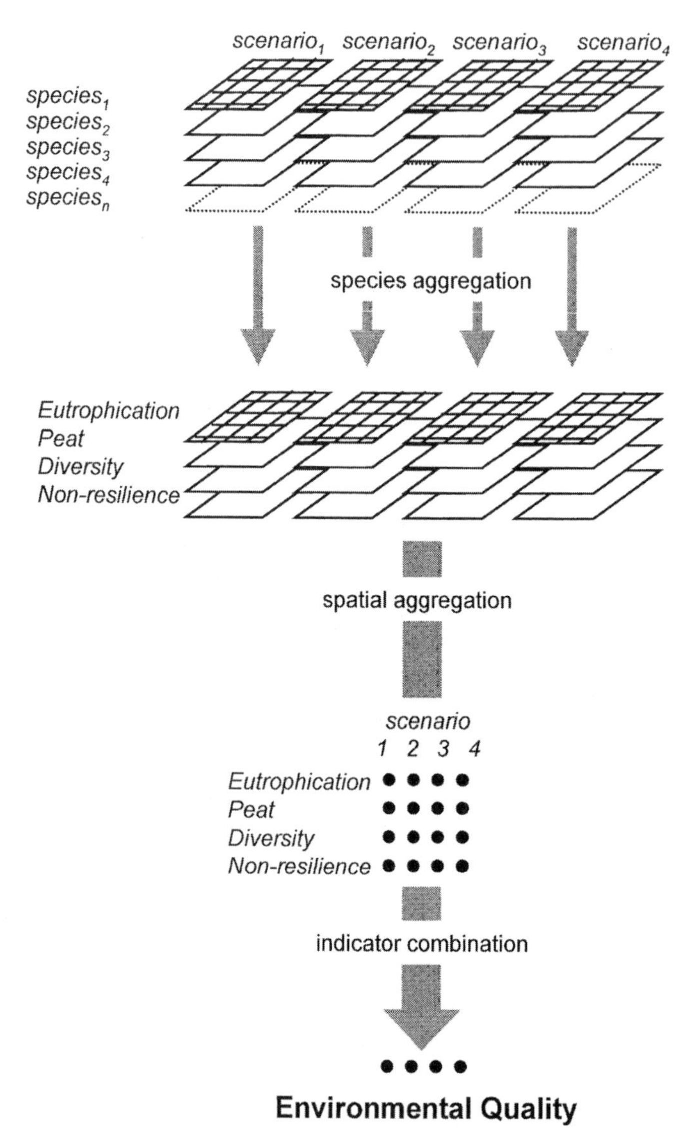

Fig. 9.4 Aggregation of model output to performance indicators and an index for environmental quality.

Step 1: identification of species

Appendix 9.1 presents the full list of species identified and classified by Dr A. Barendregt as relevant to the construction of the four indicators. Dr Barendregt not only developed the two ICHORS modules that make up the vegetation response model but also is very knowledgeable on the habitats and occurrence of plant species throughout the study area. Appendix 9.1 also indicates which of the species ultimately contribute to the calculation of the indicators.

Of the 265 (approximately) species contained in the two ICHORS modules, less than 100 actually respond to changed environmental conditions brought about by the different scenarios, regardless of the direction of the response of the species.

Step 2: standardisation of probabilities across the four scenarios

The probabilities of occurrence for each species are standardised so that changes in the probabilities of common species do not overpower those of rare species. A change from 5% to 10% is likely to be more significant than a change from 55% to 60%. The procedure involves identification of the highest probability for a species across all scenarios and all grids, setting this to 100 and recalculating the scores per grid cell for the remaining scenarios, as in the following equation:

$$S_{i,j} = \frac{100 p_{i,j}}{p_i^{\max}}, \tag{9.1}$$

where $S_{i,j}$ is the standardised probability for species i in grid cell j, $p_{i,j}$ is the probability of occurrence of species i in grid cell j and $p_i^{\max}$ is the maximum probability of occurrence of species i across all scenarios j.

Step 3: mean of standardised probabilities

Construction of each indicator involves calculation of the mean standardised probability per grid cell for those species identified in Appendix 9.1. The equation is simply:

$$I_j = \frac{\sum_{i=1}^{n} S_{i,j}}{n}, \tag{9.2}$$

where I_j is the indicator for grid cell j, $S_{i,j}$ is the standardised probability of species i occurring in grid cell j and n is the number of species.

Equation 9.2 is used for the calculation of the plant diversity indicator B. The result is positive, indicating a positive contribution to environmental quality. Its value varies between 0 (worst case) and 100 (best case). Both the eutrophic species (E) and non-resilience (R) indicators make a negative contribution to environmental quality and so the negative of the above mean is taken. The values of these indicators then vary between -100 (worst case) and 0 (best case).

The construction of the peat accumulation indicator requires an assessment of water levels before proceeding with the above calculation. Only if water levels are sufficiently high is the above calculation performed, otherwise the indicator is assigned a zero value. Further, the two ICHORS modules each contain two categories of peat-forming species:

- from ICHORS2.0, those associated with fens and those associated with bogs;
- from ICHORS3.0, those essentially aquatic and those essentially terrestrial.

These four groups of peat-forming species are, to a large extent, mutually exclusive. For example, if conditions are good for fen species, then they are bad for bog species; if a grid is surface water dominated, then it is not groundwater dominated. Consequently, the mean standardised probability is calculated for each of these four groups separately, and the maximum value is assigned to the peat accumulation indicator. In short, if $W_j > 15\,\text{cm}$, then $p_j = 0$, if not

$$P_j = \max(P_{j,k}) = \max\left(\frac{\sum_{i=1}^{n} S_{i,j,k}}{n}\right), \tag{9.3}$$

where W_j is the ground level less water level for grid cell j, P_j is the indicator for peat formation for grid cell j, $P_{j,k}$ is the mean standardised probability of species within peat-forming category k for grid cell j, $S_{i,j,k}$ is the standardised probability of species i within peat-forming category k occurring in grid cell j and n is the number of peat-forming species in category k. The peat accumulation indicator makes a positive contribution to environmental quality, with values varying between 0 (worst case) and 100 (best case).

These steps permit the aggregation across species as shown at the top of Fig. 9.4 and result in the calculation of four indicators per grid.

Step 4: spatial aggregation

The values for the four performance indicators are aggregated (mean of associated grid cells) across space to derive:

- a value of environmental quality for the region as a whole (see Fig. 9.4); and
- values of each indicator per polder (see discussion on spatial equity in Section 9.3).

The latter is straightforward, requiring the correlation of grids with polders. This can be derived from the hydrological model. With regard to the former, changes in environmental quality are not relevant for all polders in the region,

Table 9.1 *Performance indicators and index of environmental quality for the four scenarios*

Performance indicator	Reference	Agriculture	Nature	Recreation
Eutrophic species (E)	−29.15	−28.46	−29.47	−28.92
Peat formation (P)	6.27	5.94	46.97	43.09
Plant diversity (B)	23.92	23.39	25.69	24.22
Non-resilience (R)	−35.72	−34.65	−35.97	−35.90
Index of environmental quality ($= 50 + E/6 + P/6 + B/3 + R/3$)	42.25	42.49	49.49	48.47
Change relative to Reference scenario		0.24	7.24	6.22

e.g. polders under residential development. Most of the polders targeted by the scenarios, and particularly by the Nature scenario, form a corridor through the region. This corridor (part of the Blue Axis) is a recurrent theme in policy documents for restoring and strengthening the ecological infrastructure of the Netherlands. Consequently, the index for environmental quality is calculated for the corridor, and not for the region as a whole. Figure 9.5 shows the corridor and polders that contribute to this calculation.

Step 5: calculation of the index of environmental quality

Environmental quality is reflected by some mathematical combination of the above indicators. In the absence of information about this combination, the simplest case has been taken: namely, additive with equal weights for each ecosystem characteristic. Expert judgement should be assessed to test this assumption. The values of the resulting index have been adjusted so that the index varies between 0 (worst case) and +100 (best case). This leads to the following equation:

$$EQ = 50 + \frac{1}{3}\left(\frac{E+P}{2}\right) + \frac{B}{3} + \frac{R}{3}, \tag{9.4}$$

where EQ is the index of environmental quality; E, the indicator for eutrophic species, reflects ecosystem process and has a *negative* value; P, the indicator for peat accumulation, also reflects ecosystem process but has a *positive* value; B, the indicator for plant diversity, reflects ecosystem structure and has a *positive* value; and R, the indicator for non-resilience, has a *negative* value.

9.2.3 *Results and interpretation*

Table 9.1 shows the values calculated for the four performance indicators and the index of environmental quality per scenario. These results contribute to the final evaluation.

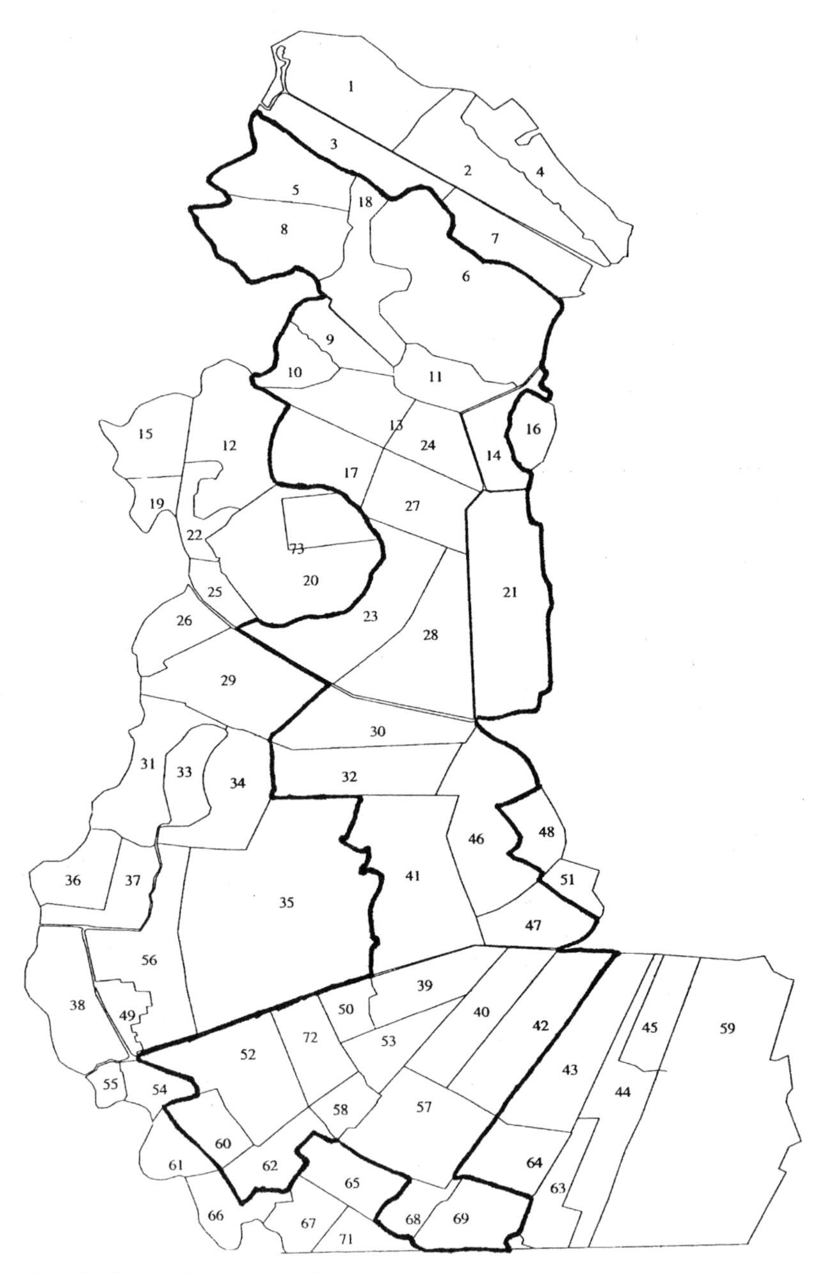

Fig. 9.5 The corridor targeted for wetlands restoration within the Vecht study area.

Of the indicators in Table 9.1, only *P* shows a clear distinction among the scenarios. The reason for this lies in the different water levels among the scenarios and the influence that this has on this indicator, rather than on differences in species probabilities. It is arguable whether differences among the remaining indicators are significant. This was surprising given that the Nature and Recreation scenarios adopted measures that were quite reasonably expected to result in substantial improvements in environmental quality. A number of possible reasons for these findings are discussed below.

1. *The scenarios may not be extreme enough to generate an improvement in environmental quality.* Additional measures for nature restoration could be needed. The measures undertaken in these scenarios are relatively passive: raising water levels and water purification. More substantial intervention to facilitate nature restoration may be needed. Such is the case for many Dutch lakes. Eutrophication and the problem of hysteresis (Hosper, 1997) are being addressed by such interventions as the dredging of sediments (to remove nutrient stocks), the culling of select fish species (notably *Abramis brama*) and restocking with macrophytes and fish species more typical of the lake ecosystem.
2. *There are errors in the models.* Both the hydrological and vegetation response models, and their output, were checked thoroughly. The conclusion from this analysis was that, while no model is perfect, these models are operating within normal parameters.
3. *The aggregation procedure – across species and across space – masks difference among the scenarios.* This is a very likely explanation. Aggregation inevitably results in loss of information. The species groups, while identified on the basis of a common factor, are still made up of species with different habitat requirements. The Vecht wetlands consist of a mosaic of different habitats and so it would not be expected that all the habitat requirements of species will be met in all places. The species used to illustrate output from the vegetation response model (*Utricularia vulgaris*) shows a clear response to the Nature scenario but only in a part of the study area. This response becomes diluted by aggregation across space. When this response is combined with that of other species, whose response may also be localised, it results in further dilution.
4. *The broad tolerance of species selected for indicator construction meant that they were not able to respond to the changes in environmental conditions created by the scenarios.* This reason would explain the small changes in the eutrophic and non-resilience indicators as these species tend to have broad tolerance of environmental conditions. However, this would not seem a likely explanation for the small changes in the biodiversity indictor.

Table 9.2 *Index of environmental quality for polders inundated in the Recreation scenario, and their adjacent polders*

Polder No.	Reference scenario	Nature scenario	Recreation scenario	Difference between Recreation and Nature
9 inundated (Heintjesrak en broeker polder noord)	−22	16	−17	
10	−19	14	14	0
11	17	16	16	0
13	−1	11	3	−8
18	−28	8	7	−1
20 inundated (Horstermeer polder)	−47	−7	−29	
12	−24	−25	−24	1
17	−10	11	−2	−13
22	−26	−11	−18	−7
23	−46	9	0	−9
25	−23	32	30	−2
27	−45	4	−5	−9

5. *The assumption that higher water levels should stimulate better environmental quality does not hold.* Analysis of the results from the hydrological model shows that higher water levels result in a greater influence of Vecht river water compared with groundwater. Water levels are sufficient, but because Vecht river water has a different chemistry, the water quality is not good enough for the return of typical fen plant species. This effect is shown in Table 9.2. Two deep polders (9 and 20) are inundated in the Recreation scenario. These polders are currently sinks, draining water from adjacent polders. Inundation would lead to less water being drained from adjacent polders. It was expected that this measure would stimulate the return of typical fen species in these adjacent polders. The Nature scenario, with higher water levels but no inundation, achieves improved environmental quality in most adjacent polders (compare the second and third columns in Table 9.2). This improvement is, to varying degrees, lost with inundation. This can be seen from the negative values in the last column in Table 9.2. If inundation is to be seriously considered as a means of improving environmental quality, this analysis suggests that additional measures are needed to increase the influence of groundwater or reduce the influence of river water. Since river water purification was included in the scenarios, reducing groundwater abstraction in the hill ridge could redress this imbalance in water sources. This measure is currently not considered in the scenarios.

9.3 Performance indicators for spatial equity

Spatial equity as an objective for the evaluation attempts to respond to the spatial (intraregional) distribution of welfare as a result of the different scenarios. The term 'welfare' is used to describe not only the economic but also the environmental state generated by the different scenarios.

There are no fixed rules for measuring conventional equity, let alone the 'spatial equity' that is being attempted here. Rose and Kverndokk (1999) provide an overview of equity within welfare economics, summarising the different perspectives and associated rules for operationalisation. Janssen and Padilla (1997) use a simple rule to include equity in an evaluation of different management choices for a mangrove forest in the Philippines. Their performance indicator for equity simply subtracts value accruing to already wealthy individuals from total value per management choice. Such a rule has limited meaning within the Netherlands and most Western nations. Rietveld (1990, 1991) analyses inequality and discusses the decomposition of inequality into an interregional and an intraregional component.

The conventional approach to measuring equity is based on the Lorenz curve and the Gini coefficient. These were considered as a means for calculating a performance indicator for spatial equity but were ultimately rejected because of their limited relevance for policy in the case study context. This study takes a very simple approach based on the key issue in equity discussions: namely, the same increase in welfare means more to poorer than to richer groups (Fig. 9.6).

The calculation of the performance indicator for spatial equity is dependent on deriving estimates of each polder's welfare within each scenario. Welfare, in this context, is defined as a combination of NPV and environmental quality. The following procedure was followed to derive welfare.

Step 1: NPV per polder The NPV calculated by the economic model captures the changes relative to the Reference scenario. The NPV for the Reference scenario can be estimated using the per hectare values presented earlier. This value captures the economic potential of the region before any changes are initiated. This NPV can be disaggregated across polders and so the economic potential of each polder, in the Reference scenario, can be represented. Based on this, the NPV values for the remaining scenarios can also be adjusted and disagreggated across polders.[2]

[2] The costs of phosphate removal were allocated equally over those polders that would be expected to benefit from improved water quality.

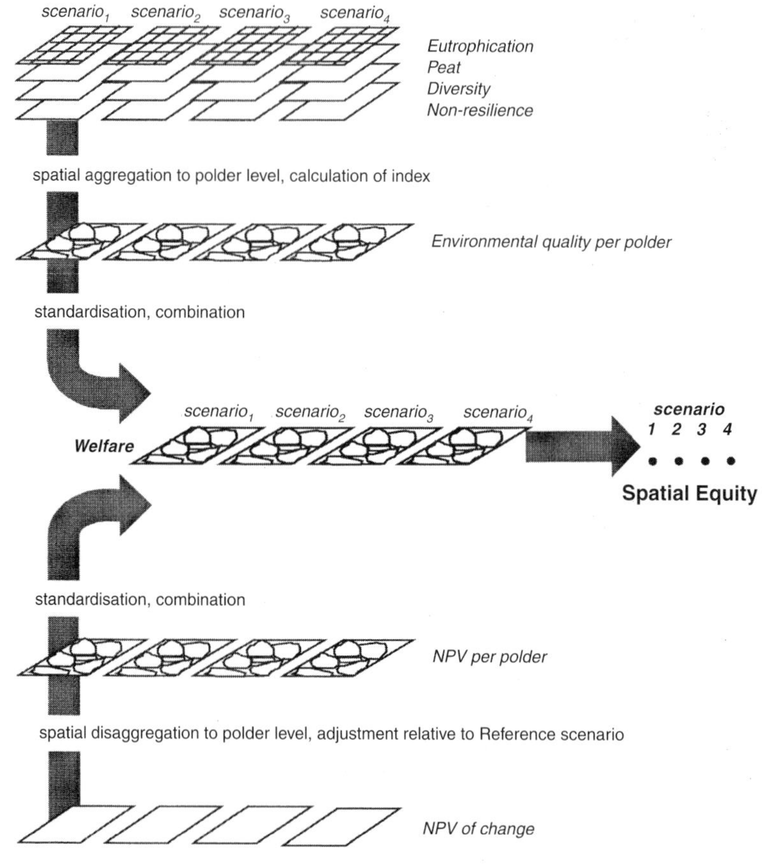

Fig. 9.6 Generation of a performance indicator for spatial equity. NPV, net present value.

Step 2: environmental quality per polder The performance indicators for environmental quality that were developed in the previous section summarise the status of the region as it is now (Reference scenario) and under three alternative scenarios. They are calculated per grid and can be aggregated either to the regional level (as done in the previous section) or to the polder level (as required here). The values per polder for economic welfare and environmental quality are presented in Appendix 9.2 (Table 9.9).

Step 3: standardisation of NPV and environmental quality Welfare per polder is some combination of economic welfare and environmental quality per polder. Since these two elements are measured in different units (monetary units and index values varying between 0 and 100), their combination must be preceded by standardisation. This is achieved separately for these two elements, by assigning the maximum across all scenarios a value of 100, and the minimum a value of 0, and recalculating the remainder. Table 9.10 in Appendix 9.2 presents the standardised values for economic welfare and environmental quality.

Step 4: calculation of welfare In the absence of better information, it is assumed that welfare is simply the sum of the (standardised) values for economic welfare and environmental quality. Other mathematical combinations are possible, as well as assigning different weights to the two elements. Table 9.11 in Appendix 9.2 presents the resulting values for welfare per polder. The unit of measure for welfare is dimensionless (index [0,100]).

The proposed procedure for measuring spatial equity tracks the changes to different classes of polder across the four scenarios. As described above, the welfare of each polder for each scenario can be calculated. These are ranked in the Reference scenario from poorest (6 for polder 53) to richest (61 for polder 35).[3] This range is then divided into four equal classes based on welfare. The classes are termed: poorest, poorer than average, richer than average, and richest. The mean welfare per polder for each class is then calculated for the same groups of polders for each scenario. The results are summarised in Table 9.3 and presented in Fig. 9.7. The following conclusions may be drawn from Fig. 9.7:

- all scenarios achieve an improvement in welfare compared with the Reference scenario, particularly for the poorest and poorer than average classes;
- the Agriculture scenario achieves the smallest improvement;
- the Nature and Recreation scenarios show substantial improvements for the poorest and poorer than average classes; and
- the Recreation scenario achieves a better improvement for the poorest class than the Nature scenario.

The performance indicator for spatial equity is calculated as follows.

1. Calculate changes in the average welfare per class relative to the Reference scenario.
2. Assign weights to these changes such that the weights add up to 1 and the greater weight is given to the poorer classes.
3. Sum the results to derive the performance indicator.

[3] Hence the need to estimate the NPV for the Reference scenario and subsequently to adjust NPV values for the remaining scenarios.

Table 9.3 *Classification of polders on the basis of welfare per polder in the Reference scenario and the average welfare per polder for each class*

Class	Polder No.[a]	Scenario			
		Reference	Agriculture	Nature	Recreation
Poorest	53, 68, 40, 72, 20, 73, 27	16.5	18.5	32.2	33.5
Poorer than average	63, 5, 52, 8, 23, 37, 64, 36, 62, 3, 18, 39, 24, 60, 22, 58, 25, 69, 67, 7, 9, 49, 28, 10, 1	27.1	27.5	38.8	39.2
Richer than average	57, 2, 21, 41, 12, 65, 32, 54, 47, 42, 15, 71, 4, 46, 13, 17, 14, 30, 26, 43, 45, 31, 11, 19, 16, 33, 50, 48	40.3	40.6	45.8	46.1
Richest	51, 29, 38, 56, 61, 55, 44, 34, 66, 59, 6, 35	52.3	52.2	54.2	56.9

[a]See Figs. 6.1 or 9.5 for the location of the polders. Note that the order of the polders in each category is from poorest to richest in the Reference scenario.

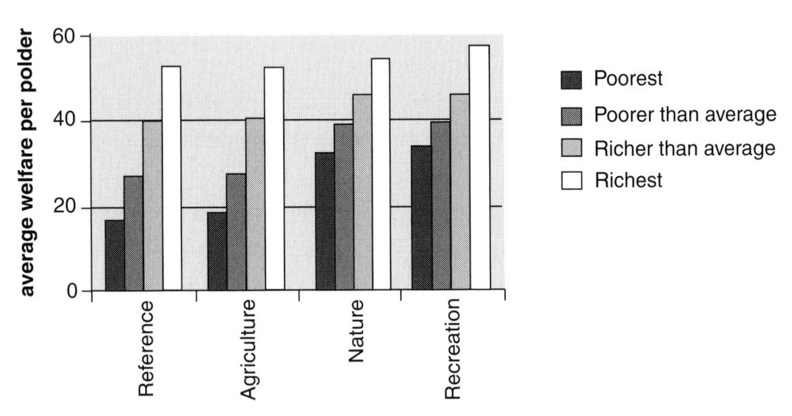

Fig. 9.7 Distribution of welfare per scenario.

Table 9.4 shows the results of these calculations. Different weights were tested but neither the ranking nor the relative positions of the scenarios changed substantially.

Close examination of Table 9.11 in Appendix 9.2 suggests that increased welfare in the Nature and Recreation scenarios derives primarily from improved environmental quality. The small increase in welfare in the Recreation scenario over the Nature scenario results from an increase in NPV and a decline in

Table 9.4 *Application of weights and subsequent calculation of the performance indicator for spatial equity*

Class	Weight	Scenario			
		Reference	Agriculture	Nature	Recreation
Poorest		0.0	2.0	15.7	17.0
	0.4	0.0	0.8	6.3	6.8
Poorer than average		0.0	0.4	11.7	12.1
	0.3	0.0	0.1	3.5	3.6
Richer than average		0.0	0.3	5.5	5.9
	0.2	0.0	0.1	1.1	1.2
Richest		0.0	−0.1	2.0	4.6
	0.1	0.0	−0.0	0.2	0.5
Summation[a]	**1.0**	**0.0**	**1.0**	**11.2**	**12.1**

[a]A weighted summation.

environmental quality. Given the apparently large increase in NPV versus the small decline in environmental quality, questions arise of scale and particularly of the role of standardisation in estimating welfare.

9.4 Conclusions

The hydrological and vegetation response models provide very detailed information across species and across space. This chapter offers an approach to condensing this output into a limited number of performance indicators. The approach is a compromise among ecological theory, the characteristics of the study area and the models being employed in the analysis.

Values for environmental quality, as calculated above, show that the Nature scenario generates the best environmental quality. Recreation comes a close second. However, differences among the four scenarios are small, suggesting limited sensitivity to the different scenarios. Two reasons among the different explanations for this seem most likely. First, aggregation across species and space leads to the loss of information. The species groups, while identified on the basis of a common factor, still comprise species with different habitat requirements. It is not to be expected that all habitat requirements of species will be met in all places. Second, the raising of water levels may have the undesirable effect of increasing the influence of Vecht river water relative to groundwater. The poor quality of this water may be limiting the response of plant species.

Values for spatial equity, as calculated above, show that the Recreation scenario generates the most improvement to the welfare of poorer polders. Nature comes a close second. Given the large difference between these scenarios with regard to NPV, and the small difference with regard to environmental quality, questions arise concerning the procedure for combining these two elements. Chapter 10 presents an overall evaluation of the analysis.

Appendix 9.1 Full list of species identified

Tables 9.5–9.8 list the plant species used to derive the performance indicators in the two ICHORS modules.

Table 9.5 *Plant species identified and used in the construction of the eutrophic species indicator (E)*[a]

ICHORS2.0	ICHORS3.0
Angelica sylvestris	***Acorus calamus***
Calamagrostis canescens[b]	*Alisma plantago aquatica*
Carex acutiformis[b]	*Alopecurus geniculatus*
Carex paniculata	*Bidens cernua*
Carex riparia[b]	*Bidens connata*
Epilobium hirsutum	*Calamagrostis canescens*
Eupatorium cannabinum	***Carex acuta***
Festuca arundinacea	***Carex acutiformis***
Juncus effusus	*Carex paniculata*
Lycopus europaeus	*Carex riparia*
Lysimachia vulgaris	*Catabrosa aquatica*
Lythrum salicaria	*Ceratophyllum demersum*
Phalaris arundinacea	***Elodea nutalii***
Polygonum amphibium	***Epilobium hirsutum***
Salix cinerea	*Epilobium parviflorum*
Solanum dulcamara	*Glyceria maxima*
Stellaria uliginosa	*Iris pseudacorus*
	Lemna gibba
	Lemna minor
	Phalaris arundinacea
	Polygonum amphibium
	Polygonum hydropiper
	Potamogeton pectinatus
	Potamogeton pusillus
	Rorippa amphibia
	Solanum dulcamara
	Stellaria uliginosa
	Typha angustifolia
	Zannichellia palustris

[a]Species identified by A. Barendregt; those used in construction of the indicator are given in bold.
[b]Also as dominant form.

Table 9.6 *Plant species identified and used in the construction of the peat accumulation indicator* (P)[a]

ICHORS2.0		ICHORS3.0	
Bog species	Fen species	Aquatic species	Terrestrial species
Carex curta	***Alnus glutinosa***	***Carex pseudocyperus***	***Acorus calamus***
Carex rostrata	*Calamagrostis canescens*	*Cicuta virosa*	*Berula erecta*
Drosera rotundifolia	*Carex acutiformis*[b]	*Elodea canadensis*	***Carex acutiformis***
Erica tetralix	***Carex paniculata***	***Equisetum fluviatile***	*Carex disticha*
Eriophorum angustifolium[c]	*Solanum dulcamare*	***Sagittaria sagittifolia***	*Thelypteris palustris*
	Thelypteris palustris[c]	*Stratiotes aloides*	*Typha angustifolia*
	Typha angustifolia	***Utricularia vulgaris***	*Veronica beccabunga*
Polytrichum comm.	*Carex riparia*[c]	*Hottonia palustris*	*Carex paniculata*
Potentilla erecta	*Iris pseudacorus*	***Hydrocharis morsus ranae***	***Juncus subnodulosus***
Sphagnum fimbriatum	*Lysimachia vulgaris*	*Nuphar lutea*	***Lysimachia thyrsiflora***
Sphagnum flexuosum	***Phragmites australis***	*Oenanthe aquatica*	*Peucedanum palustre*
Sphagnum palustre	*Rumex hydrolapathum*	***Potamogeton natans***	*Phragmites australis*
Sphagnum squarrosum	*Salix cinerea*	*Potamogeton obtusifolius*	*Potentilla palustris*

[a]Species identified by A. Barendregt; those used in construction of the indicator are given in bold.

[b]Also as dominant form.

[c]Only as dominant form.

Table 9.7 *Plant species identified and used in the construction of the biodiversity indicator (B)*[a]

ICHORS2.0			ICHORS3.0		
Calthion vegetation	Florally rich, herbaceous	Quagmire	Clear water vegetation	Typical fen vegetation	Bank-fen vegetations
Caltha palustris	*Angelica sylvestris*	*Carex diandra*	*Butomus umbellatus*	***Achillea ptarmica***	***Acorus calamus***
Cardamine pratensis	*Calamagrostis canescens*[b]	*Carex flacca*	***Elodea nutalii***	*Cardamine flexuosa*	*Calamagrostis canescens*
Carex disticha	*Carex acuta*	*Carex lasiocarpa*	*Hydrocotyle vulgaris*	***Carex pseudocyperus***	*Caltha palustris*
Carex palustris	*Epilobium hirsutum*	*Carex panicea*	***Lysimachia nummularia***	*Cicuta virosa*	***Carex acutiformis***
Dactylorhiza majalis	*Epilobium palustre*	*Carex pseudocyperus*	*Mentha aquatica*	*Elodea canadensis*	*Carex paniculata*
Eleocharis palustrisp	*Eupatorium cannabinum*	*Carex rostrata*	*Myosotis palustris*	***Equisetum fluviatile***	*Carex riparia*
Equisetum fluviatile	***Filipendula ulmaria***	*Equisetum fluviatile,*	*Nymphoides peltata*	*Hottonia palustris*	*Cirsium palustre*
Galium palustre	***Holcus lanatus***[b]	*Galium uliginosum*	***Potamogeton natans***	***Hydrocharis morsus ranae***	*Eupatorium cannabinum*
Hypericum tetrapterum	*Iris pseudacorus*	***Juncus subnodulosus***[b]	*Potamogeton pusillus*	*Lemna trisulca*	***Filipendula ulmaria***
Juncus articulatus	*Juncus effusus*	*Lysimachia thyrsiflora*	*Ranunculus circinatus*	*Oenanthe aquatica*	*Hypericum tetrapterum*
Juncus acutiflorus	*Lythrum salicaria*	*Menyanthes trifoliata*	***Sagittaria sagittifolia***	*Potamogeton obtusifolius*	*Iris pseudacorus*
Lotus uliginosus	*Solanum dulcamara*	*Pedicularis palustris*		***Potamogeton trichoides***	*Juncus articulatus*
Lychnis flos cuculi	*Stachys palustris*	*Peucedanum palustre*		*Potentilla palustris*	***Juncus conglomeratus***
Lycopus europaeus	***Symphytum officinale***	***Potentilla palustris***		*Ranunculus flammula*	***Juncus subnodulosus***

Table 9.7 (cont.)

ICHORS2.0			ICHORS3.0		
Calthion vegetation	Florally rich, herbaceous	Quagmire	Clear water vegetation	Typical fen vegetation	Bank-fen vegetations
Lysimachia vulgaris		***Ranunculus lingua***		*Stratiotes aloides*	*Lotus uliginosus*
Myosotis palustris		*Utricularia minor*		***Utricularia vulgaris***	***Lychnis flos cuculi***
Rhinanthus serotinus				*Veronica beccabunga*	*Lycopus europaeus*
Scutellaria galericulata				*Veronica catenata*	***Lysimachia thyrsiflora***
Valeriana officinalis					***Lysimachia vulgaris***
					Phragmites australis
					Potamogeton lucens
					Rumex hydrolapathum
					Scipus lacustris lacustris
					Scutellaria galericulata
					Solanum dulcamara
					Thelypteris palustris
					Typha angustifolia

[a]Species identified by A. Barendregt; those used in construction of the indicator are given in bold.
[b]Also as dominant form.

Table 9.8 *Plant species identified and used in the construction of the non-resilience indicator (R)*[a]

ICHORS2.0 (reed-sedge vegetation)	ICHORS3.0
Carex acutiformis[b]	***Bidens cernua***
Carex paniculata	*Bidens connata*
Carex riparia[b]	*Ceratophyllum demersum*
Festuca arundinacea	***Epilobium hirsutum***
Glyceria maxima	*Epilobium tetragonum*
Mentha aquatica	*Glyceria maxima*
Phalaris arundinacea	*Juncus bufonis*
Phragmites australis	***Lemna gibba***
Polygonum amphibium	*Nuphar lutea*
Sparganium erectum	*Nymphaea alba*
Thelypteris palustris	*Phalaris arundinacea*
Typha angustifolia	*Polygonum amphibium*
	Polygonum hudropiper
	Spirodela polyrhiza

[a] Species identified by A. Barendregt; those used in construction of the indicator are given in bold.
[b] Also as dominant form.

Appendix 9.2 Environmental quality and economic welfare indices on a polder level

Tables 9.9–9.11 list the environmental quality index and economic welfare for each polder.

Table 9.9 *Environmental quality (index, [0,100]) and economic welfare (million €) per polder for the four scenarios*

Polder No.[a]	Reference		Agriculture		Nature		Recreation	
	EQ	NPV	EQ	NPV	EQ	NPV	EQ	NPV
1	42.7	23.9	44.3	32.4	42.7	23.9	42.7	16.3
2	43.4	22.1	42.6	28.2	43.4	22.1	43.3	22.1
3	39.6	3.6	39.6	8.5	39.6	3.6	39.6	−4.9
4	45.2	45.2	45.2	43.9	45.2	45.2	45.2	35.9
5	37.6	4.5	37.8	10.6	47.6	−30.4	46.8	37.8
6	53.3	72.0	52.9	70.2	53.6	71.3	53.6	64.9
7	41.8	2.7	41.6	6.4	50.3	−39.2	49.7	52.6
8	38.2	5.9	38.2	13.8	47.1	−39.5	46.7	58.2
9	41.8	2.3	41.8	5.3	49.1	−32.7	43.1	43.2
10	42.5	2.3	42.6	5.3	48.6	−32.7	48.9	35.3
11	48.5	3.2	48.5	0.0	48.5	3.2	48.5	3.2
12	41.6	97.9	41.6	96.5	41.4	97.9	41.5	194.4
13	45.4	52.8	45.2	53.4	48.1	49.3	46.1	57.0
14	46.7	2.7	46.0	3.2	53.8	−39.2	53.1	41.2
15	45.4	19.8	45.6	22.9	45.5	19.8	45.5	19.8
16	48.8	0.0	48.8	0.0	48.8	0.0	48.8	−18.5
17	44.9	70.2	48.6	70.2	49.4	70.2	46.5	59.4
18	40.1	3.6	40.4	4.3	47.1	−52.3	47.1	69.4
19	48.6	0.9	48.7	2.1	48.4	0.9	48.4	−4.1
20	36.1	7.7	36.1	18.1	43.7	−111.2	40.4	144.8
21	42.7	50.2	42.3	43.4	49.6	−3.4	48.5	109.7
22	41.0	0.5	41.0	0.0	44.4	0.5	42.7	−0.3
23	36.3	81.8	37.1	82.3	47.4	56.1	45.5	105.1
24	40.4	11.0	39.8	14.0	50.0	−6.5	49.8	26.0
25	41.1	1.4	40.8	3.2	51.9	−10.3	51.4	11.5
26	47.3	2.7	47.3	6.4	47.1	1.6	47.1	−3.0
27	36.1	21.0	39.2	25.9	45.7	−8.1	44.0	52.1
28	41.6	38.5	41.5	39.1	51.1	10.5	50.1	70.3
29	47.2	140.3	47.2	140.3	52.5	140.3	51.8	271.9
30	45.4	63.0	45.4	61.1	52.1	35.0	52.5	96.2
31	46.6	71.5	46.5	73.3	53.0	61.0	52.3	76.2
32	43.4	53.6	45.4	53.7	51.5	46.6	49.0	61.9
33	48.8	26.3	48.8	26.3	48.8	26.3	48.8	26.3
34	49.4	96.5	49.4	96.5	49.4	96.5	49.4	96.1
35	45.9	429.9	45.8	429.9	45.9	429.9	45.7	846.5
36	38.8	2.7	38.8	6.4	38.8	2.7	38.8	−8.7
37	38.5	2.3	38.5	5.3	38.5	2.3	38.5	−7.5
38	50.3	22.4	50.3	29.1	50.3	22.4	50.3	14.0

Table 9.9 (cont.)

Polder No.	Reference		Agriculture		Nature		Recreation	
	EQ	NPV	EQ	NPV	EQ	NPV	EQ	NPV
39	40.1	4.1	39.4	4.8	49.5	−59.5	49.9	78.5
40	35.3	21.1	34.9	21.7	46.0	−35.5	45.3	87.2
41	42.4	67.3	42.6	68.3	50.2	21.9	48.9	121.4
42	45.1	31.0	44.0	37.7	52.2	−8.1	50.8	68.9
43	47.4	6.4	47.5	14.9	47.9	5.7	47.9	−3.6
44	51.4	12.4	51.9	17.3	51.7	11.7	51.7	12.4
45	47.9	1.8	46.8	4.3	48.1	1.1	48.1	0.7
46	46.0	13.7	46.1	20.4	53.7	−24.8	52.3	59.4
47	44.8	3.2	45.0	7.4	53.7	−45.8	52.3	61.4
48	49.7	0.0	50.2	0.0	50.2	0.0	50.2	−17.6
49	41.5	18.0	41.5	18.6	51.6	14.5	50.1	19.3
50	48.8	26.3	48.7	26.3	49.0	25.6	49.0	26.3
51	50.2	9.6	50.2	8.7	50.2	9.6	50.2	9.6
52	37.1	31.8	37.1	32.7	46.2	−52.1	45.8	129.0
53	30.2	1.8	33.6	4.3	41.9	−12.9	38.8	5.1
54	44.7	1.4	44.5	3.2	44.7	1.4	44.7	0.6
55	51.6	0.5	51.6	1.1	51.6	0.5	51.6	0.0
56	47.8	122.8	47.7	122.8	53.0	122.8	52.9	243.5
57	42.6	39.2	42.7	44.7	49.7	−24.4	49.2	113.6
58	40.7	18.0	41.7	18.6	47.8	13.8	46.5	21.7
59	48.9	174.4	48.9	165.3	48.9	174.4	48.9	174.0
60	40.6	10.1	42.4	12.0	48.8	−1.0	47.2	22.1
61	51.1	0.0	51.1	0.0	51.1	0.0	51.1	0.0
62	38.1	35.1	39.3	35.1	47.7	34.4	46.5	35.1
63	37.0	11.5	37.3	15.2	37.2	10.8	37.2	11.1
64	38.2	19.4	36.5	21.8	36.8	18.7	36.8	16.0
65	43.4	26.3	43.6	26.3	43.2	25.6	43.1	22.0
66	51.9	1.4	51.7	3.2	51.9	1.4	51.9	1.4
67	41.5	2.7	42.3	6.4	41.4	2.7	41.2	2.7
68	35.3	1.8	35.5	4.3	45.6	−12.9	44.9	9.9
69	41.0	20.1	39.8	23.7	47.2	−1.6	45.4	33.3
71	45.7	8.8	46.3	8.8	46.4	8.8	46.4	8.8
72	35.7	11.5	35.8	12.0	44.8	−30.5	44.5	61.4
73	36.4	2.7	36.5	0.0	36.1	2.7	36.1	2.7

EQ, environmental quality index; NPV, net present value (million €).

[a]Polder 70 was not part of the analysis.

Table 9.10 *Standardised economic welfare and environmental quality [0,100] per polder for the four scenarios*

Polder No.[a]	Reference		Agriculture		Nature		Recreation	
	EQ	NPV	EQ	NPV	EQ	NPV	EQ	NPV
1	53.0	14.1	59.9	15.0	53.0	14.1	53.0	13.3
2	55.7	13.9	52.5	14.5	55.8	13.9	55.6	13.9
3	39.9	12.0	40.0	12.5	39.9	12.0	39.9	11.1
4	63.6	16.3	63.5	16.2	63.6	16.3	63.6	15.4
5	31.4	12.1	32.0	12.7	73.6	8.4	70.5	15.6
6	97.9	19.1	96.3	18.9	99.3	19.1	99.2	18.4
7	49.1	11.9	48.1	12.3	85.1	7.5	82.7	17.1
8	33.9	12.2	33.9	13.0	71.6	7.5	70.0	17.7
9	49.3	11.8	49.3	12.2	80.1	8.2	54.8	16.1
10	52.2	11.8	52.3	12.2	78.0	8.2	79.4	15.3
11	77.5	11.9	77.6	11.6	77.5	11.9	77.5	11.9
12	48.3	21.8	48.4	21.7	47.4	21.8	47.8	31.9
13	64.2	17.1	63.7	17.2	75.7	16.8	67.4	17.6
14	69.7	11.9	66.9	11.9	100.0	7.5	97.1	15.9
15	64.6	13.7	65.1	14.0	64.9	13.7	64.9	13.7
16	78.8	11.6	78.9	11.6	78.7	11.6	78.7	9.7
17	62.5	18.9	77.8	18.9	81.2	18.9	69.0	17.8
18	42.1	12.0	43.2	12.1	71.8	6.1	71.4	18.9
19	77.8	11.7	78.3	11.8	77.2	11.7	77.1	11.2
20	24.7	12.4	24.8	13.5	57.2	0.0	43.2	26.7
21	53.1	16.9	51.4	16.1	82.3	11.3	77.5	23.1
22	45.6	11.7	45.6	11.6	60.0	11.7	52.9	11.6
23	26.0	20.1	29.1	20.2	72.7	17.5	64.6	22.6
24	43.2	12.8	40.8	13.1	83.8	10.9	82.9	14.3
25	46.2	11.7	44.7	11.9	92.1	10.5	89.7	12.8
26	72.5	11.9	72.5	12.3	71.6	11.8	71.4	11.3
27	25.1	13.8	38.0	14.3	65.6	10.8	58.3	17.1
28	48.3	15.6	47.7	15.7	88.4	12.7	84.3	18.9
29	72.0	26.3	72.0	26.3	94.5	26.3	91.6	40.0
30	64.2	18.2	64.2	18.0	92.8	15.3	94.3	21.7
31	69.4	19.1	69.2	19.3	96.5	18.0	93.8	19.6
32	55.7	17.2	64.6	17.2	90.2	16.5	79.6	18.1
33	78.8	14.4	78.8	14.4	78.8	14.4	78.8	14.4
34	81.4	21.7	81.4	21.7	81.4	21.7	81.4	21.6
35	66.5	56.5	66.0	56.5	66.4	56.5	65.7	100.0
36	36.3	11.9	36.6	12.3	36.3	11.9	36.3	10.7
37	35.1	11.8	35.1	12.2	35.1	11.8	35.1	10.8
38	85.1	13.9	85.1	14.6	85.1	13.9	85.1	13.1

Table 9.10 (cont.)

Polder No.	Reference		Agriculture		Nature		Recreation	
	EQ	NPV	EQ	NPV	EQ	NPV	EQ	NPV
39	42.0	12.0	39.1	12.1	81.8	5.4	83.4	19.8
40	21.4	13.8	19.7	13.9	67.0	7.9	64.0	20.7
41	51.5	18.6	52.3	18.7	84.8	13.9	79.4	24.3
42	63.1	14.8	58.3	15.5	93.3	10.8	87.3	18.8
43	73.0	12.3	73.3	13.2	75.1	12.2	75.1	11.2
44	89.8	12.9	91.9	13.4	91.1	12.8	91.1	12.9
45	75.1	11.8	70.4	12.1	75.8	11.7	75.9	11.7
46	67.1	13.0	67.3	13.7	99.7	9.0	93.4	17.8
47	61.8	11.9	62.8	12.4	99.5	6.8	93.4	18.0
48	82.5	11.6	84.6	11.6	84.9	11.6	84.9	9.8
49	47.7	13.5	47.7	13.5	90.8	13.1	84.3	13.6
50	78.8	14.4	78.5	14.4	79.4	14.3	79.5	14.4
51	84.8	12.6	84.7	12.5	84.7	12.6	84.7	12.6
52	29.2	14.9	29.2	15.0	67.9	6.2	66.2	25.1
53	0.0	11.8	14.4	12.1	49.4	10.3	36.5	12.1
54	61.5	11.7	60.6	11.9	61.5	11.7	61.5	11.7
55	90.7	11.7	90.7	11.7	90.7	11.7	90.7	11.6
56	74.7	24.4	74.3	24.4	96.5	24.4	96.3	37.0
57	52.3	15.7	52.8	16.3	82.7	9.1	80.5	23.5
58	44.4	13.5	48.6	13.5	74.5	13.0	69.0	13.9
59	79.2	29.8	79.2	28.9	79.2	29.8	79.2	29.8
60	43.9	12.7	51.5	12.9	78.8	11.5	72.0	13.9
61	88.6	11.6	88.3	11.6	88.6	11.6	88.6	11.6
62	33.3	15.3	38.4	15.3	74.2	15.2	69.2	15.3
63	28.8	12.8	29.9	13.2	29.6	12.7	29.7	12.8
64	33.6	13.6	26.4	13.9	28.1	13.6	28.1	13.3
65	55.9	14.4	56.9	14.4	55.2	14.3	54.7	13.9
66	91.9	11.7	91.2	11.9	92.0	11.7	91.9	11.7
67	47.8	11.9	51.2	12.3	47.5	11.9	46.6	11.9
68	21.4	11.8	22.4	12.1	65.4	10.3	62.0	12.6
69	45.7	13.7	40.7	14.1	72.1	11.4	64.3	15.1
71	65.8	12.5	68.3	12.5	68.4	12.5	68.5	12.5
72	23.4	12.8	23.5	12.9	61.8	8.4	60.5	18.0
73	26.4	11.9	26.5	11.6	25.1	11.9	25.1	11.9

EQ, environmental quality index; NPV, net present value (million €).
[a]Polder 70 was not part of the analysis.

Table 9.11 *Welfare per polder for the four scenarios (index [1,100])*

Polder No.[a]	Scenario			
	Reference	Agriculture	Nature	Recreation
1	33.6	37.4	33.6	33.2
2	34.8	33.5	34.9	34.8
3	26.0	26.2	26.0	25.5
4	40.0	39.8	40.0	39.5
5	21.8	22.3	41.0	43.0
6	58.5	57.6	59.2	58.8
7	30.5	30.2	46.3	49.9
8	23.1	23.5	39.5	43.8
9	30.6	30.7	44.1	35.4
10	32.0	32.3	43.1	47.3
11	44.7	44.6	44.7	44.7
12	35.1	35.0	34.6	39.9
13	40.7	40.4	46.2	42.5
14	40.8	39.4	53.8	56.5
15	39.1	39.5	39.3	39.3
16	45.2	45.3	45.2	44.2
17	40.7	48.4	50.1	43.4
18	27.0	27.6	38.9	45.1
19	44.7	45.1	44.5	44.1
20	18.6	19.1	28.6	35.0
21	35.0	33.8	46.8	50.3
22	28.6	28.6	35.8	32.2
23	23.1	24.7	45.1	43.6
24	28.0	26.9	47.4	48.6
25	29.0	28.3	51.3	51.3
26	42.2	42.4	41.7	41.3
27	19.5	26.1	38.2	37.7
28	32.0	31.7	50.5	51.6
29	49.1	49.1	60.4	65.8
30	41.2	41.1	54.0	58.0
31	44.2	44.2	57.3	56.7
32	36.5	40.9	53.3	48.8
33	46.6	46.6	46.6	46.6
34	51.6	51.6	51.6	51.5
35	61.5	61.2	61.5	82.9
36	24.1	24.4	24.1	23.5
37	23.5	23.6	23.5	23.0
38	49.5	49.9	49.5	49.1
39	27.0	25.6	43.6	51.6

Table 9.11 (cont.)

Polder No.	Scenario			
	Reference	Agriculture	Nature	Recreation
40	17.6	16.8	37.4	42.3
41	35.1	35.5	49.3	51.8
42	39.0	36.9	52.0	53.1
43	42.6	43.2	43.6	43.1
44	51.3	52.6	52.0	52.0
45	43.5	41.2	43.8	43.8
46	40.1	40.5	54.4	55.6
47	36.9	37.6	53.2	55.7
48	47.1	48.1	48.2	47.3
49	30.6	30.6	51.9	49.0
50	46.6	46.4	46.9	46.9
51	48.7	48.6	48.6	48.6
52	22.0	22.1	37.0	45.6
53	5.9	13.2	29.8	24.3
54	36.6	36.3	36.6	36.6
55	51.2	51.2	51.2	51.2
56	49.6	49.4	60.4	66.7
57	34.0	34.6	45.9	52.0
58	28.9	31.1	43.8	41.4
59	54.5	54.1	54.5	54.5
60	28.3	32.2	45.1	43.0
61	50.1	50.0	50.1	50.1
62	24.3	26.9	44.7	42.2
63	20.8	21.5	21.2	21.3
64	23.6	20.2	20.8	20.7
65	35.1	35.6	34.7	34.3
66	51.8	51.6	51.9	51.8
67	29.8	31.7	29.7	29.3
68	16.6	17.2	37.8	37.3
69	29.7	27.4	41.8	39.7
71	39.2	40.4	40.5	40.5
72	18.1	18.2	35.1	39.2
73	19.2	19.0	18.5	18.5

[a]Polder 70 was not part of the analysis.

10

Evaluation of the scenarios

10.1 Introduction

The performance indicators developed in Chapter 9 reflect three objectives (net present value (NPV), environmental quality and spatial equity), which will be used to evaluate the scenarios. This section presents the results of this evaluation. Its aim is to rank the scenarios, to identify where trade-offs among objectives may be needed, to identify spatial features within the ranking, and to identify elements within the scenarios that bear further consideration in regional planning. A distinction is made between point evaluation (the subject of Section 10.2) and spatial evaluation (the subject of Section 10.3). This distinction is illustrated below in Fig. 10.5.

10.2 Point evaluation

The point evaluation stems from the performance indicators calculated for the region as a whole. Table 10.1 presents the results of calculating performance indicators, as outlined in the previous section. This table forms the basis for the evaluation.

The objective function for this evaluation is some mathematical combination of the three objectives appearing in Table 10.1. The form of this function is unknown, and it is further complicated by the use of different measurement units for each objective. Taken individually, each objective yields a different ranking of the scenarios, where > indicates 'preferred to':

NPV: Recreation > Agriculture> Reference > Nature;
environmental quality: Nature > Recreation > Agriculture > Reference;
spatial equity: Recreation > Nature > Agriculture > Reference.

Table 10.1 *Performance indicators for the three evaluation objectives per scenario: the effects table for the evaluation*

Objective	Performance indicator	Scenario			
		Reference	Agriculture[a]	Nature[a]	Recreation[a]
Net present value	NPV (million €)	0	175.6	−226.4	1982.6
Environmental quality	Index [0,100]	0	0.24	7.24	6.22
Spatial equity	Index [−100,100]	0.0	1.0	11.2	12.1

[a]Values represent change relative to the Reference scenario.

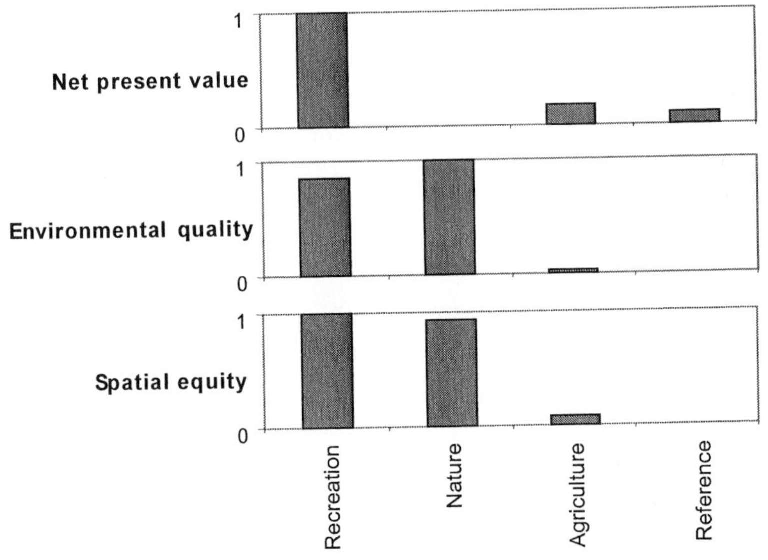

Fig. 10.1 Graphical presentation of the effects table and ranking of scenarios. The scores for each objective have been standardised between 0 and 1.

Techniques derived from multicriteria analysis, as implemented in the decision-support system DEFINITE (e.g. Janssen, 1992; Janssen and van Herwijnen, 1994), were used to derive an overall ranking and to analyse trade-offs between objectives. The results are presented below.

Figure 10.1 presents graphically the information from Table 10.1, standardised such that all performance indicators range between 0 and 1.

Assuming an additive objective function and equal weights, as given in the following vector of weights:

$$\begin{pmatrix} \omega_{NPV} \\ \omega_{EQ} \\ \omega_{SE} \end{pmatrix} = \begin{pmatrix} 1/3 \\ 1/3 \\ 1/3 \end{pmatrix}, \tag{10.1}$$

(where EQ is environmental quality and SE is spatial equity) the ranking is:

Recreation > Nature > Agriculture > Reference.

The Recreation scenario scores best on two objectives and well on the third and so is clearly the preferred scenario for this set of weights. Assigning different weights to the objectives can reverse the ranking of the Recreation and Nature scenarios. Considering the weight of each objective separately, rank reversal occurs under the following sets of weights:

- $\omega_{NPV} = 0.032$ (the remaining weight of 0.968 is allocated equally across the other two objectives); and
- $\omega_{EQ} = 0.796$ (the remaining weight of 0.204 is allocated equally across the other two objectives).

No rank reversal can be achieved by changing ω_{SE}.

This reversal can also be achieved by changing the weights assigned to all three objectives simultaneously. The weights closest to the original weights, but producing rank reversal (as calculated by DEFINITE) are given by the following vector:

$$\begin{pmatrix} \omega_{NPV} \\ \omega_{EQ} \\ \omega_{SE} \end{pmatrix} = \begin{pmatrix} 0.042 \\ 0.524 \\ 0.434 \end{pmatrix} \tag{10.2}$$

The value of ω_{NPV} must be very small for rank reversal. While probably unlikely, this issue has not been checked with stakeholders in the region. The Recreation scenario dominates both the Reference and Agriculture scenarios, as no combination of weights would lead to either of these scenarios being ranked first.

The conclusion from the overall ranking is that Recreation is the preferred scenario.

Trade-offs and the degree of conflict between any two objectives can become more transparent by using scatter diagrams, such as in Fig. 10.2. Each diagram plots the scores for the four scenarios on two objectives. Scores are standardised between 0 (the worst scenario) and 100 (the best scenario). Consider the scatter diagram in Fig. 10.2, in which NPV is plotted against environmental quality. The scenario with the highest environmental quality (Nature) can be found at the top of the diagram and the scenario with the highest NPV (Recreation) can be found on the right. The ideal scenario for these two objectives would combine optimal performance on NPV with optimal performance on quality. This ideal scenario would score 100 for both objectives and would be found in the upper-right corner of the diagram. It is clear from Fig. 10.2 that such an ideal scenario does not exist, although Recreation is a reasonable approximation. The remaining two scenarios perform poorly on one or both objectives.

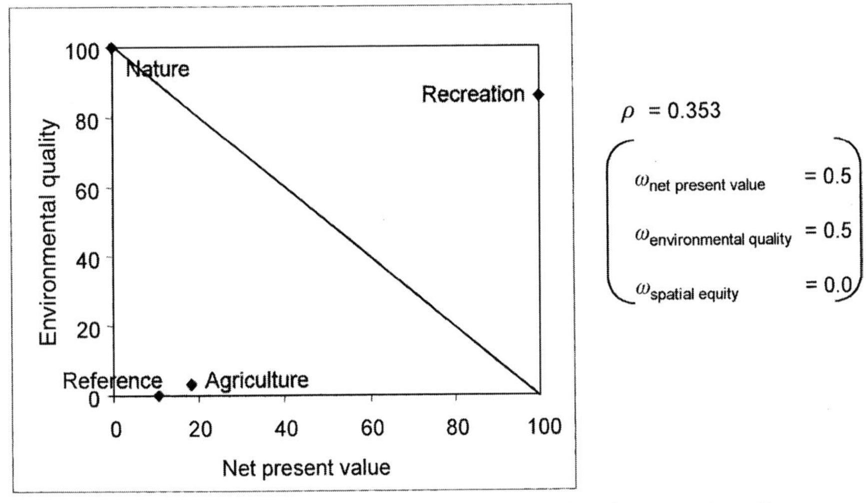

Fig. 10.2 Trade-off between net present value and environmental quality.

The level of conflict between these two objectives is reflected by the correlation coefficient (ρ).[1] A negative value indicates conflict between objectives, with increases in one objective being paired with decreases in the other. The positive value (0.353) for the coefficient indicates no conflict between these objectives.

The line shown in Fig. 10.2 can be used to rank the alternatives visually. The angle of the line implies that equal weight is given to NPV and environmental quality. All points on this line have the same distance from the ideal point at (100,100).[2] Moving this line from top right to bottom left ranks the scenarios. The first to cross the line and, therefore, the best scenario is Recreation. The ranking is:

Recreation > Nature > Agriculture > Reference.

A change in relative weights of the two objectives is reflected by a change in the angle of the line. Only by assigning a weight smaller than 0.124 (compared with 0.5; see Fig. 10.2) to NPV would Recreation no longer be the preferred scenario. No combination of weights would ever lead to either Reference or Agriculture being the preferred scenario: both are dominated by Recreation. This shows that the ranking is not very sensitive to the weights assigned to the objectives.

[1] Addition or removal of scenarios may influence the relative position of the remaining scenarios and will also influence the correlation coefficient.

[2] Distance is defined here as the sum of the distance along the x-axis and the distance along the y-axis. Since the line intersects the two axes at the same distance from the ideal point, all points on the line share the same distance from the ideal point.

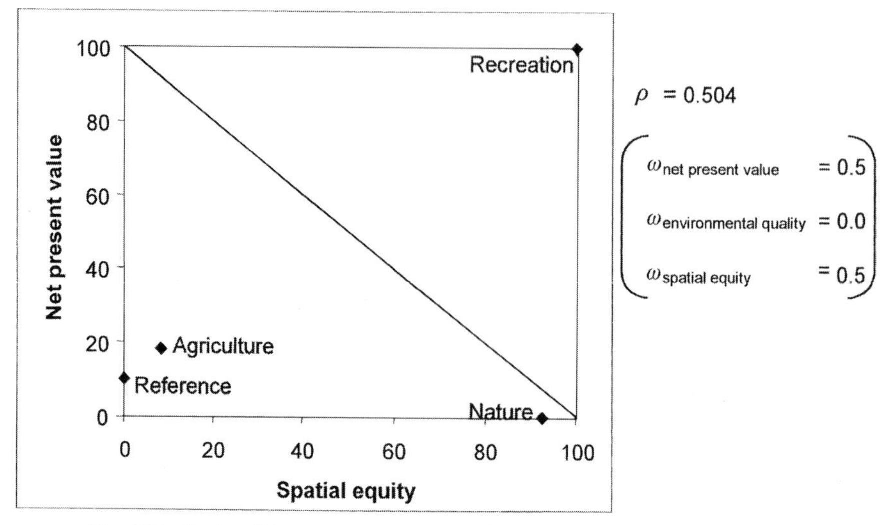

Fig. 10.3 Trade-off between net present value and spatial equity.

Figure 10.3 presents the scatter diagram for spatial equity versus NPV. Recreation scores ideally. The ranking in this case, with equal weighting, is the same as with Fig. 10.2:

Recreation > Nature > Agriculture > Reference.

Because Recreation dominates the other scenarios, rank reversal between the preferred and second most preferred scenarios cannot occur. The correlation coefficient (0.504) indicates no conflict between these objectives.

Figure 10.4 presents the scatter diagram for environmental quality versus spatial equity. Nature and Recreation scenarios lie close to, but not on, the ideal point. These two scenarios dominate the Agriculture and Reference scenarios. The ranking in this case, with equal weighting, is:

Nature > Recreation > Agriculture > Reference.

The ranking of the Nature and Recreation scenarios is sensitive to weighting. If the weight assigned to environmental quality is reduced to less than 0.347 (compared with 0.5), then Recreation becomes the preferred scenario. The correlation coefficient (0.986) shows that there is no conflict between these two objectives. The above analysis would lead to the following conclusions.

- Recreation is the preferred scenario, scoring best on two of the three objectives.
- Only by assigning NPV a weight smaller than 0.0424 (e.g. 1/3) would this be overturned so that Nature becomes the preferred scenario.

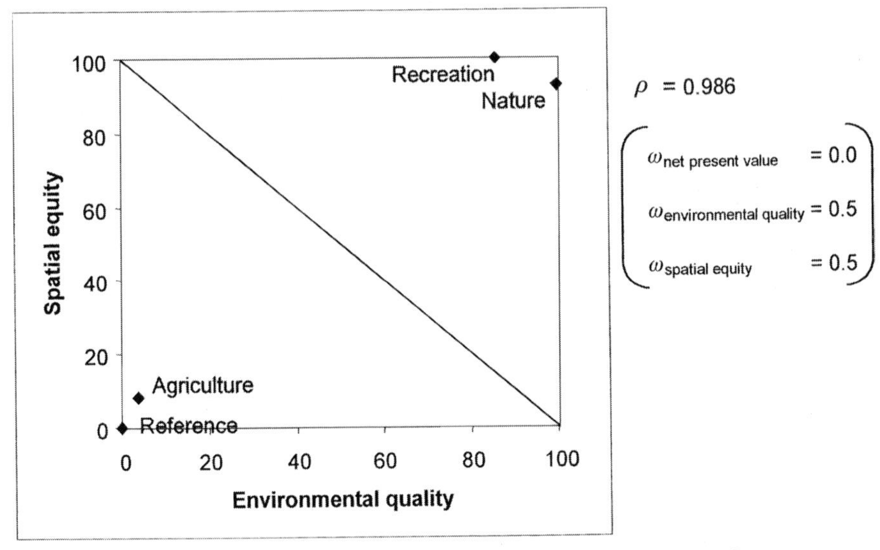

Fig. 10.4 Trade-off between environmental quality and spatial equity.

- Recreation dominates the Agriculture and Reference scenarios. If the environmental quality objective were excluded, Recreation would also dominate the Nature scenario.
- There is no conflict between the different pairs of objectives. This contrasts with the results of Janssen and Padilla (1997), Lorenz *et al.* (2001a) and Veeren and Lorenz (2002) where economic efficiency conflicted with both environmental quality and equity. This lack of conflict is grounds for optimism that a scenario could be devised in which all three objectives score ideally.

10.3 Spatial evaluation

As discussed in Chapter 9, the generation of performance indicators for environmental quality requires aggregation of spatially detailed model output and is likely to result in the loss of information. Techniques have been developed for spatial evaluation (van Herwijnen, 1999). These techniques attempt to retain the spatial pattern that different scenarios achieve. Van Herwijnen has shown that this may lead to a different ranking of alternatives. This section tests whether spatial evaluation would lead to conclusions different from those made in Section 10.2. Only two objectives are considered in this section. Spatial equity is already an amalgam of spatial characteristics and, unlike NPV and environmental quality, cannot be disaggregated to the polder level.

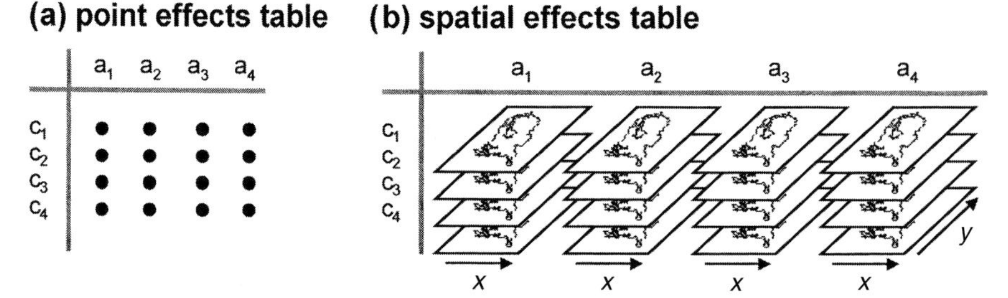

Fig. 10.5 Spatial effects and point effects tables. (From van Herwijnen, 1999.)

The models used in this analysis predict the effects of alternative scenarios per polder. In the case of the hydrological and vegetation response models, environmental effects can be predicted at an even smaller scale, per 500 m × 500 m grid cell. This means that the effect of a scenario for two evaluation objectives – NPV and environmental quality – is not just one value but a matrix of values that can be visualised as a map. Instead of an effects table presenting one value per objective for the whole region (as used for these two objectives in the evaluation above, and as illustrated in Fig. 10.5a), a table of maps may be derived in which the spatial detail is maintained. The spatial effects table in Fig. 10.5b shows the effects from four scenarios (a_1, a_2, a_3 and a_4) for four different objectives (c_1, c_2, c_3 and c_4). The essential difference between these two tables is that the point effects table is two dimensional while the spatial effects table is four dimensional. This table contains scenarios (a_i), objectives (c_i) and the coordinates that specify the geographical location of the objectives' scores (x_i,y_i).

The presentation of each matrix as a map can support decision making. Unfortunately, reading maps is often not as easy as it seems. Comparing sets of maps visually gets exponentially more difficult if more than two maps are compared at the same time (van Herwijnen and Rietveld, 1999; see also Bertin, 1981; Kraak and Ormeling, 1996; MacEachren, 1994; Monmonier, 1991). A procedure developed by van Herwijnen (1999) and designed to help the decision maker to structure and simplify the spatial performance of alternatives has been used in this case study. The procedure is summarised in Fig. 10.6 and discussed further below. The starting point for a spatial evaluation is the four-dimensional spatial effects table (Fig. 10.5b) which, for the Vecht case study, comprises four scenarios (a_i), performance indicators for environmental quality and NPV (c_i) and coordinates giving their location (x_i,y_i). Aggregation of the four-dimensional effects table to derive a ranking of the alternatives can follow two paths, as shown by the block arrows in Fig. 10.6. The two paths differ according to the order of the two substeps.

Table 10.2 *Derivation of a single score per scenario according to path 1 in Fig. 10.6*

Standardised scores	Scenario			
	Reference	Agriculture	Nature	Recreation
Net present value	0.10	0.18	0.00	1.00
Environmental quality	0.00	0.03	1.00	0.86
Combined	0.10	0.21	1.00	1.86

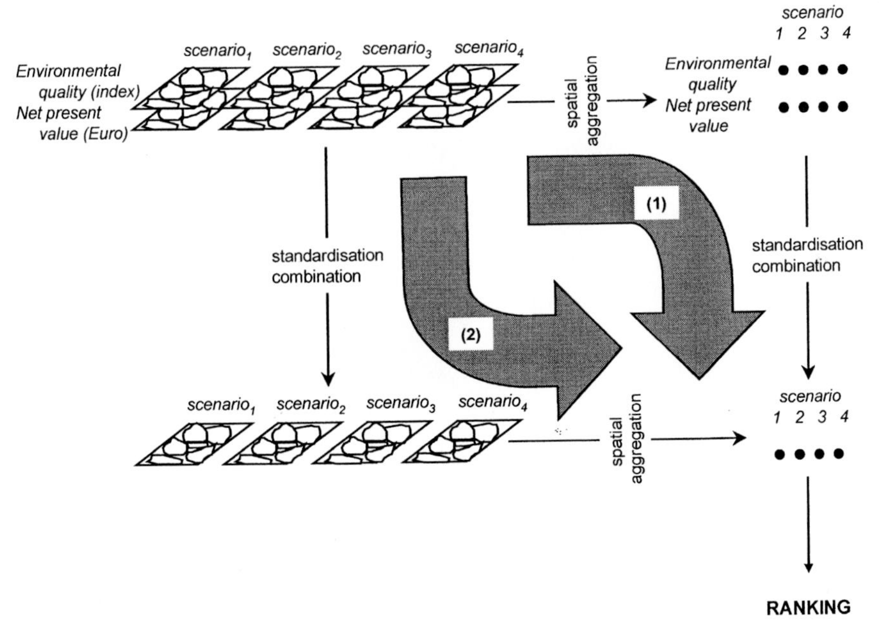

Fig. 10.6 Alternative paths for evaluation of scenarios according to objectives with a spatial character.

1. Aggregation across space to derive a single value for each objective per scenario (and thereby the conventional two-dimensional effects table) and then standardising and combining objectives to derive a single value per alternative.
2. Standardising and combining objectives to derive a single value for each location per alternative (a map per alternative) and then aggregating across space to derive a single value per alternative.

Van Herwijnen (1999) has shown that these paths can yield different rankings of alternatives. The first path above would lead to the results presented in Table 10.2. This table is derived from Table 10.1 by standardising scores for the

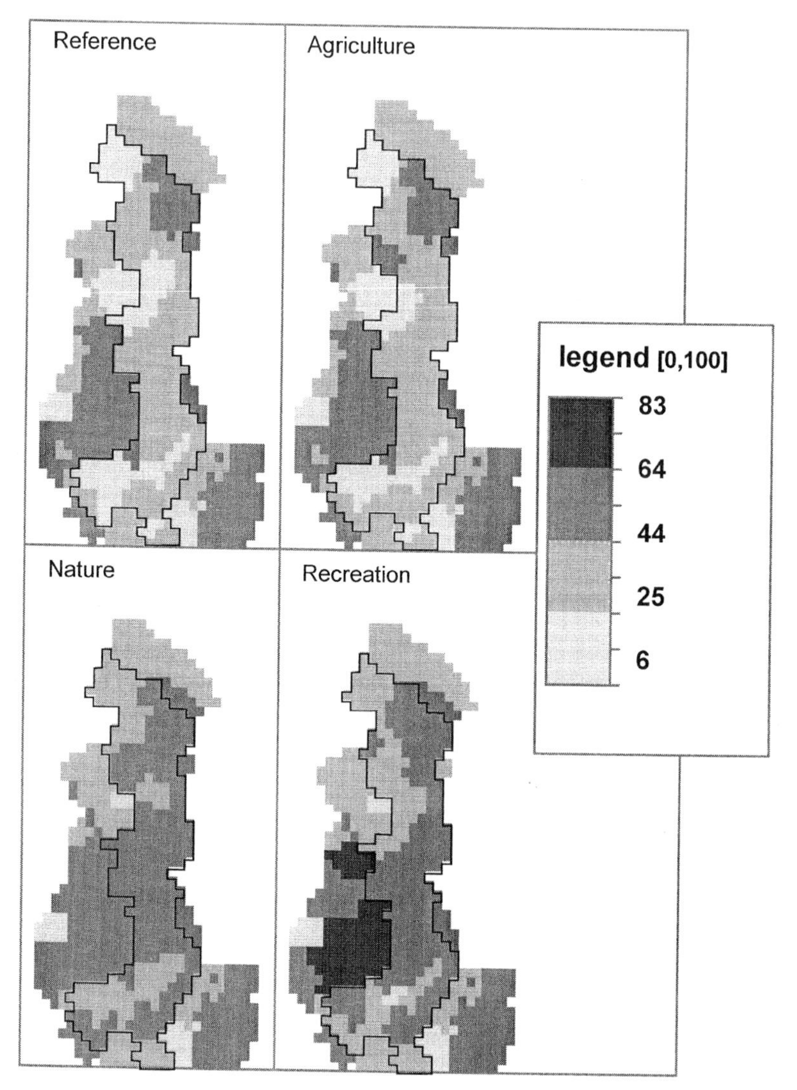

Fig. 10.7 Welfare per polder.

two objectives between 0 and 1. The combination rule is, again, the simplest: namely, additive with equal weights. The ranking is clearly:

Recreation > Nature > Agriculture > Reference.

The procedure for standardisation and combination within the second path in Fig. 10.6 is the same as that described in Chapter 8 to derive welfare per polder. The resulting set of maps for each scenario is presented in Fig. 10.7. Note that the corridor up the middle of the region (see also discussion in Ch. 9) targeted

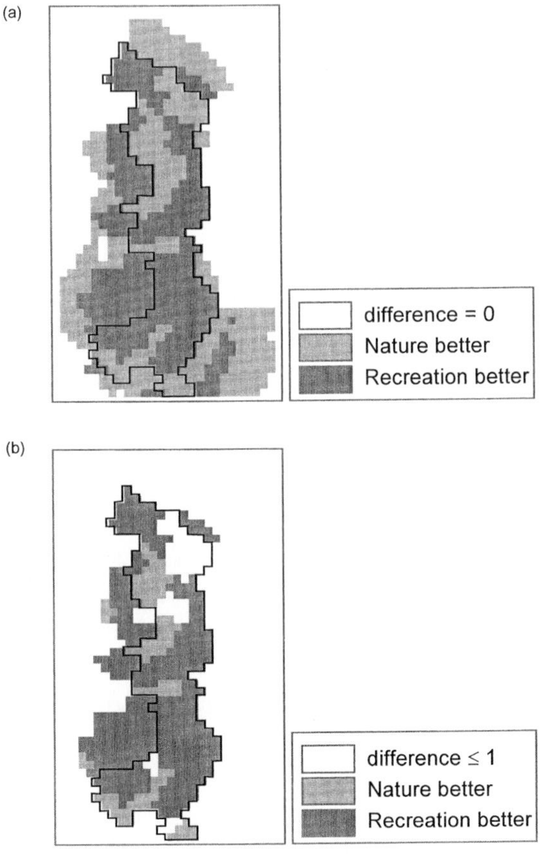

Fig. 10.8 Difference map for welfare, comparing the Nature and the Recreation scenarios.

for nature restoration is also shown in these maps. The most preferred scenario should be 'darkest'. It is clear from Fig. 10.7 that the Nature and Recreation scenarios are substantially better than the Reference and Agriculture scenarios. Which is better, Recreation or Nature, is not clear.

Separation of these two scenarios was attempted by testing the degree of difference in the results for the two scenarios. Figure 10.8 presents difference maps for welfare. These maps consider only the Recreation and Nature scenarios and show which of the two scores best for each polder. White areas denote polders where the two scenarios score equally (Fig. 10.8a) or where the difference between the two is less than, or equal to, one unit of welfare (Fig. 10.8b). Many of the polders where Nature is the better scenario are lost in Fig. 10.8b, showing that the difference between these scenarios is very small. However, no polders with Recreation as the better scenario are lost in this figure. This suggests that Recreation is often a close second where Nature is the better scenario, while the converse does not hold. The conclusion would be that, overall, Recreation was the better scenario.

However, this conclusion is less certain if only the corridor is considered. The majority of polders have a higher welfare in the Recreation scenario (18 polders) than in the Nature scenario (14 polders). Of the 14 polders where Nature is the better scenario in Fig. 10.8a, 10 remain with Nature as the better scenario in Fig. 10.8b. However, as shown in Table 10.3, which presents welfare for the corridor polder in the two scenarios, the two scenarios do not differ greatly and the mean welfare per polder, for these corridor polders, differs by less than one unit of welfare.

The ranking of scenarios as a result of following path 2 in Fig. 10.6 would be (excluding the Agriculture and Reference scenarios):

Recreation > Nature.

While this ranking agrees with that derived from path 1 of Fig. 10.6, there is a larger degree of uncertainty associated with it than was suggested by the earlier evaluations. Maintaining spatial detail in the evaluation, by following path 2 of Fig. 10.6, has exposed this uncertainty, which affects the corridor running up the middle of the study area. This corridor was explicitly targeted by the scenarios for nature restoration in both scenarios, and for capturing the benefits of this restoration via recreation. In generating an income from nature restoration, it would have been reasonable to expect that the Recreation scenario would score substantially better than the Nature scenario. The most likely reason for the poor separation of these two scenarios lies with the inundation of two polders in the Recreation scenario. As discussed in Chapter 9, inundation constrained improvements in environmental quality.

A second reason could lie with the standardisation procedure behind the calculation of welfare. As has also been discussed in Chapter 9, the scenarios do not differ substantially in their environmental quality. These small differences become amplified in the standardisation procedure, in which the smallest polder value is given a zero value and the largest a value of 100. Conversely, the very large differences in NPV, particularly between the Nature and Recreation scenarios, are compressed by the same procedure. Large differences in NPV lead to small differences in welfare, whereas small differences in environmental quality lead to large differences in welfare.

10.4 Conclusions

The point evaluation clearly shows that Recreation is the most preferred scenario and that Nature is the second most preferred scenario. Only if the

Table 10.3 *Welfare per corridor polder for the Nature and Recreation scenarios (index [1,100])*

Polder No.	Nature	Recreation
3	26.0	25.5
5	41.0	43.0
6	59.2	58.8
8	39.5	43.8
9	44.1	35.4
10	43.1	47.3
11	44.7	44.7
13	46.2	42.5
14	53.8	56.5
17	50.1	43.4
18	38.9	45.1
21	46.8	50.3
23	45.1	43.6
27	38.2	37.7
28	50.5	51.6
30	54.0	58.0
32	53.3	48.8
39	43.6	51.6
40	37.4	42.3
41	49.3	51.8
42	52.0	53.1
46	54.4	55.6
47	53.2	55.7
50	46.9	46.9
52	37.0	45.6
53	29.8	24.3
57	45.9	52.0
58	43.8	41.4
60	45.1	43.0
62	44.7	42.2
68	37.8	37.3
69	41.8	39.7
72	35.1	39.2
Mean	44.6	45.4

weight given to NPV is substantially reduced is this ranking reversed. Recreation dominates the remaining scenarios. The spatial evaluation confirms this ranking but exposes a greater degree of uncertainty in the order of the Nature and Recreation scenarios, particularly for the corridor that was targeted for nature restoration.

The difference between the Nature and Recreation scenarios, apart from the stimulation of recreation, is explained by the inundation of two polders. The results of the analysis suggest that inundation achieves less improvement in environmental quality than smaller increases in water level.

Obviously, the goal of further scenario development would lie with both improving environmental quality and capturing the benefits of this nature restoration by stimulating recreation. Two qualifications to seeking such a scenario are needed. First, the analysis has been limited to plant species and has not attempted to capture the effect of nature restoration on higher order species, such as birds. Second, this analysis has also not considered the possible adverse impacts of increased recreation on environmental quality, for example trampling of vegetation, increased litter, noise and other disturbance.

The index for environmental quality does not differ substantially across the four scenarios. It is questionable whether these differences are significant and, therefore, whether their amplification as a result of the various standardisation procedures is legitimate. The uncertainty that is exposed by the spatial evaluation could be the result of this amplification. Excluding environmental quality from the point evaluation would lead to an evaluation based on two objectives – NPV and spatial equity – in which case, Recreation would dominate.

11

Conclusions: policy and research implications

11.1 Purpose

Wetland areas are under heavy pressure and require the attention of both natural and social scientists. In order to arrive at useful policy and management suggestions to ameliorate problems related to the human use of, and impacts on, wetlands, the integration of theories, models and indicators of relevant natural and social sciences is necessary. For this purpose, the present study has developed an integrated research method that combines hydrology, ecology and economics, using spatial modelling and evaluation techniques. This method was applied to the Vecht wetlands area (De Vechtstreek) in the centre of the Netherlands, an area of about 160 km^2 in size that includes natural and artificial lakes, reedbeds, fens, peatlands, marshes, forests and wet pasturelands. The area's land cover, land use and landscape are the product of an interaction between human activities and natural processes over centuries. The main activities are agriculture and recreation, while the main conflicts are between agriculture, on the one hand, and nature conservation and outdoor recreation, on the other hand. Much information is available for the area, via natural science studies performed in the past, as well as through local, regional and national policy documents.

11.2 Method of integrated research

The method of integrated study entailed the heuristic integration of hydrological quality and quantity modelling, vegetation response modelling and economic modelling and accounting. The inputs and outputs of the integrated approach had an explicit spatial dimension. The hydrological–ecological part of the modelling and analysis was performed on a grid basis (500 m × 500 m) and subsequently aggregated to the level of polders of, on average, 200 ha, to make it consistent with the economic spatial disaggregation basis. This approach follows from the idea that real integration of hydrological, ecological and economic information can best proceed via the concrete interfacing of models in the context of spatial scenarios. The integration was realised at three levels. A first integration level involved the formulation of development scenarios for the Vecht area that include consistent settings for the hydrological and economic parameters, while also taking into account the feasibility of changes to land uses and vegetation cover. A second integration level was explicit feedback between the natural science and the economic models. This was based on the inclusion of nutrients from agriculture in the water-quality modelling and the influence of environmental quality on recreation. A third integration level concerned the aggregation of information in three performance indicators reflecting economic efficiency (i.e. net present value (NPV)), environmental quality and spatial equity, which were combined in order to study conflicting objectives and trade-offs, as well as to rank the scenarios. Finally, integration also occurred within the natural science context, as water-quality, water-quantity and vegetation-response models were linked in a heuristic sequence.

Three alternative development scenarios have been studied: the stimulation of agriculture, of nature, and of recreation. These scenarios describe alternative spatial hydrological and economic conditions for the Vecht, such as land use, water levels, intensity of agriculture, management of nature reserves, recreational services and investment in phosphate-removal plants. The scenarios define settings at the polder level, thus taking account of the spatial dimension of hydrological and economic parameters.

The results under a particular scenario are calculated as changes relative to a Reference scenario, which is defined as a continuation of the present situation (in terms of water levels, land uses, etc.). Detailed spatial information about the impacts of scenarios on polders is generated, relating to employment, nutrient run-off, generation of income, costs of restoration, and ecological productivity, diversity and resilience of the wetlands.

11.3 Natural science modelling and results

The natural science modelling linked three models.

1. A water quantity model describing amounts of groundwater and surface water expected at specified locations in the present hydrological conditions.

2. A water quality model, describing the chemistry of groundwater and surface water at specified locations.
3. An ecological model describing the presence of 265 wetland plant species at specified locations, through probabilities of encountering species that depend on local environmental (abiotic) conditions.

A set of boundary conditions is needed to define the domains of the models. Some of these boundary conditions can be manipulated in scenarios; others are static in the time domain adopted here (e.g. as soil texture).

The natural science modelling gave rise to the following insights. Environmental conditions necessary for the stimulation of characteristic wetland species in the Vecht area are high water tables; water with low concentrations of chloride, magnesium, sulphate, sodium, potassium and nutrients (phosphorus, nitrogen, ammonia); and buffered conditions, notably high pH values caused by high concentrations of calcium and bicarbonate. Two opposing influences act on water quality and the presence of the desired fen and peatland species. First, drainage attracts fresh groundwater from surrounding areas, which positively affects the wetland species. Second, maintaining high water tables inevitably means inflow of water from the river Vecht, which is rich in nutrients and has relatively high salinity; this creates negative indirect effects on the wetland species. The net effect of these processes is also influenced by drainage; acidification resulting from the input of rainwater; drainage of peatlands, with subsequent mineralisation of peat and release of additional nutrients; and intensification of agriculture with extra input of manure and fertilisers. As a result of these – partly conflicting – processes, stimulating nature will only be possible in specific parts of the region.

11.4 Economic analysis and results

The economic analysis consisted of a spatial–economic model and a spatial cost–benefit analysis (CBA) procedure. The model describes agriculture, nature conservation and recreation, as well as some additional activities. The inputs to the spatial–economic model include the scenario settings, area-based economic and agricultural–environmental data, and environmental quality indicators per polder. The outputs of the model include environmental indicators of run-off and surplus of nitrogen and phosphate for use in the hydrological model, discounted cost and benefit indicators for use in the evaluation procedure, and other economic indicators such as changes in employment. The accounting procedure follows a financial CBA that is restricted to financial transactions. At first sight, an economic CBA might seem more relevant for the purpose of the present study, as it also includes the benefits of improved environmental quality and nature, which fall outside the realm of the market. However, given the approach followed in the study, these non-market value effects can be captured directly via indicators that measure the improvements in environmental quality. Moreover, including both these indicators and the non-market values would create a problem of double-counting. Although adapting the weights of the various indicators in the evaluation procedure could technically solve this, it would create additional uncertainty in the evaluation procedure.

Under the Agriculture scenario, farm management is intensified in various polders. Intensive agriculture is associated with more dairy cows per hectare. In the Nature scenario, parts of the agricultural land are converted into nature areas. The costs of this conversion are the acquiring, restructuring and maintenance costs, and the costs of foregone benefits of agriculture (i.e. the opportunity costs). These opportunity costs are the current revenues of agriculture and the cost of installing, at three locations, phosphate-removal plants to extract phosphorus from the water that flows into the Vecht area. The Recreation scenario describes lands that are converted into nature areas and designated for recreation, and polders that are flooded for the purpose of water recreation. The financial benefits and costs of outdoor recreation are estimated using available data on spending on recreation and revenues of recreation-related industries. The scenarios are mutually consistent, in the sense that the benefits per hectare of agricultural land are used as opportunity costs in the Nature and Recreation scenarios.

Two additional elements have been examined. First, the impact on benefits of possible feedback from biodiversity and nature quality to outdoor recreation has been calculated, where it is assumed that a higher environmental quality increases the benefits per hectare of land open for recreation, because improved natural surroundings will attract more recreationists. Second, for illustrative purposes, benefit transfer of use and non-use values for nature was performed on the basis of a valuation study for a Dutch case that seemed best comparable with the Vecht area. Although both elements lead to quantitative monetary indicators, they have not been incorporated in the overall evaluation (for the reasons outlined above).

The results of the economic accounting can be summarised as follows. Intensification of agriculture has a positive effect on the NPV (total discounted benefits minus total discounted costs). Under the Nature scenario, the costs of converting agricultural land are the only financial indicators on which the calculation of the NPV is based. Therefore, the NPV is negative (i.e. there are costs but no financial benefits). The NPV under the Recreation scenario is positive: the increase in recreational benefits outweighs the costs of converting the land use. Some of the results have been summarised for the subareas 'north', 'middle', 'south' of the Vecht area. The economic changes are most significant in the north. It should be noted, however, that these conclusions do not take into account the benefits of nature conservation or improved environmental quality.

11.5 Indicators, spatial evaluation and scenario ranking

The evaluation procedure involved a point and a spatial evaluation to derive an overall ranking of the scenarios. This was done on the basis of three objectives: NPV, spatial equity, and environmental quality. Much attention was given to the development of indicators for the last two objectives. At a theoretical level, the indicators of environmental quality attempt to capture three ecosystem characteristics (process, structure and resilience). Plant species were aggregated into four groups reflecting eutrophication and peat formation (ecosystem process), plant diversity (ecosystem structure) and non-resilience (resilience). No other information was available about higher-order species, and the assessment of environmental quality was limited to plant species (and to the probability of their occurrence). Spatial equity was approached from the perspective that the same increase in welfare means more to poorer than to richer polders. Since many other approaches to spatial equity can be defended, the results are merely illustrative of what an analysis of spatial equity in a multidimensional indicator context entails.

The scenarios, particularly Nature and Recreation, focused on stimulating the return of wetland vegetation in a corridor that runs from north to south through the study area. The results show that this was achieved, although the degree of improvement to nature was somewhat limited. There are two likely explanations for this. First, indicators for environmental quality require aggregation across species and across space. This aggregation could be diluting real improvements to wetland quality. Second, raising water levels does not necessarily mean improving conditions for wetlands, because of the hydrological balance between nutrient-rich Vecht river water and nutrient-poor groundwater. The influence of groundwater can be increased only by further reducing groundwater abstraction in the hill ridge. This factor was not incorporated in the scenarios.

The objective of the evaluation was to rank the scenarios, assess the sensitivity of this ranking and identify where trade-offs between objectives may be needed. In terms of the most-preferred scenario and assuming an additive objective function, the results of the evaluation were quite clear: Recreation is the most-preferred scenario, followed by Nature. Recreation scores best on NPV and on spatial equity, and second on environmental quality. Nature becomes the most preferred scenario only if the weight assigned to NPV is substantially reduced. Nature scores best on environmental quality, second on spatial equity and worst on NPV. No combination of weights leads to either the Agriculture or the Reference scenario being the most preferred. Both are dominated by Recreation.

Care should be taken with the conclusion that Recreation is the most preferred scenario. No negative feedback from increased recreational use to environmental quality has been considered. Examples of such adverse impacts include trampling, litter and noise. Species that are not represented in the ecological model, such as birds and mammals, are more likely to be sensitive to these adverse effects and to large numbers of recreationists.

A pair-wise comparison of objectives showed that the level of conflict between pairs of objectives is small. The Recreation scenario, in scoring well on all three objectives, demonstrates the small level of conflict.

The point evaluation suggested that the ranking was not very sensitive. However, the spatial evaluation showed that this sensitivity was much greater if the spatial patterns of the economic and environmental objectives were considered, and in particular within the corridor targeted for nature development. The scenarios did not achieve substantial differences in environmental quality and it is questionable whether these differences are significant. The standardisation procedure within the evaluation magnifies these small differences. The sensitivity between the Nature and Recreation scenarios, which the spatial evaluation suggested, could in fact be an artefact of standardisation. Even so, the spatial evaluation offers the possibility of avoiding the loss of information that aggregation can cause. Therefore (additional) explicit spatial evaluation is very important in the case of a multifunctional area like the Vecht area. It is too large to be treated as a hydrologically, ecologically or economically homogeneous area. In other words, spatial differences matter.

11.6 Further research

This study presented an approach that can best be regarded as a compromise among ecological theory, features of the study area and the knowledge and models available to the participants in the study. As a result, various improvements are conceivable. More research is possible on details of the various elements of the method of integrated research adopted in this study. For example, the different types of intensive and extensive agriculture may require more attention. Given the average or aggregate treatment of agriculture on a polder level, it is not possible to identify agriculture settings with particular types of agriculture, such as organic farming. Moreover, more specific details could be included. This would require an assessment of the profitability and environmental effects of different types of agriculture in a detailed manner, that is, moving towards a production function approach. There could also be further work on specifying the set-up costs of scenarios, which have only been approximately covered here. For instance, the assessment of costs of arranging suitable conditions for stimulating recreation can be improved. It would also be interesting to look at other scenarios, such as futurist ones, in which various technological solutions are examined. These may include water-quality improvement technology, or stimulating multifunctionality and capturing nature values via the creative integration of infrastructure, housing, recreation and nature. In terms of evaluation, other spatial equity criteria and aggregation or objective functions can be considered. Next, attention may be given to providing more detailed links between the natural science and the economic models. This may include developing other types of model, such as dynamic systems models, in which concepts like 'resilience' could be operationalised in a more satisfactory way. The most ambitious approach in this sense would be to realise a fully integrated spatial hydrological–ecological–economic model, that is, one without any heuristic linkages. Finally, the study has offered a quite unique range of aggregation and disaggregation levels. This has served to illustrate where the aggregation of indicators leads to significant information losses, implying difficulties in terms of ranking or evaluation and interpretation. The final conclusion is that the search for performance indicators has to continue.

References

Adam, P. (1998). Australian saltmarshes: a review. In A. J. McComb and J. A. Davis (eds.) *Wetlands for the Future*. Adelaide: Gleneagles Publishing, pp. 287–295.

Adriaanse, A. (1993). *Environmental policy performance indicators*. Amsterdam: SDU.

Adriaanse, M., J. van der Kraats, P. G. Stoks and R. C. Ward (eds.) (1994). *Proceedings of the International Workshop Monitoring Tailor-made I*, 20–23 September 1994, Beekbergen, the Netherlands.

Aelmans, F. G. (1985). *Grondwaterkaart van Nederland. [Groundwater Map of the Netherlands.]* Lelystad (20 west), Harderwijk (26 west en oost).

Alcamo, J. (ed.) (1994). *Image 2.0: Integrated Modelling of Global Climate Change*. Dordrecht: Kluwer.

Alcamo, J., R. Shaw and L. Hordijk (eds.) (1990). *The Rains Model of Acidification: Science and Strategies in Europe*. Dordrecht: Kluwer.

Allen, P. M. and J. M. McGlade (1987). Modelling complex human systems: a fisheries example. *European Journal of Operational Research* **30**, 147–167.

Anderson, M. P. and W. W. Woesner (1992). *Applied Groundwater Modelling. Simulation of Flow and Advective Transport*. New York: Academic Press.

Andrewartha, H. G. (1972). *Introduction to the Study of Animal Populations*. Chicago, IL: University of Chicago Press.

Anon. (1997). *The Ramsar Convention Manual: A Guide to the Convention on Wetlands* (Ramsar, Iran, 1971). Appendix 8: *Ramsar Classification System for Wetland Type*, 2nd edn. Gland, Switzerland: Ramsar Convention Bureau. http://www.ramsar.org/lib_manual_1.htm.

(2001). *The Ramsar Convention on Wetlands: 26th Meeting of the Ramsar Standing Committee – Agenda Papers*. 3–7 December 2001, Gland, Switzerland. http://www.ramsar.org/key_sc26_docs_cop8_07.htm.

Antuma S. J. F., P. B. M. Berentsen and G. W. J. Giesen (1994). *Naar een Duurzame Landbouw in Noord-Brabant: Kosten en Baten voor Boer en Milieu*. Wageningen: Landbouwuniversiteit Wageningen.

Arntzen, J. W. (1989). Environmental Pressure and Adoptation in Rural Botswana. Ph.D. Thesis, Free University, Amsterdam.

Aronsson, T. and K.-G. Löfgren (1999). Renewable resources: forestry. In J. C. J. M. van den Bergh (ed.) *Handbook of Environmental and Resource Economics*. Cheltenham, UK: Edward Elgar, pp. 122–140.

Arrow, K. J., R. Solow, P. Portney, E. Leamer, R. Radner and H. Schuman (1993). *Report of NOAA Panel on Contingent Valuation*. [58 Fed. Reg. 46.] Washington, DC: National Oceanic and Atmospheric Administration.

Ayres, R. U. and A. V. Kneese (1969). Production, consumption and externalities. *American Economic Review* **59**, 282–297.

Ayres, R. U., K. J. Button and P. Nijkamp (1999). *Global Aspects of the Environment*, Vols. 1 and 2. Cheltenham, UK: Edward Elgar.

Ayres, R. U., J. C. J. M. van den Bergh and J. M. Gowdy (2001). Weak versus strong sustainability: economics, natural sciences and consilience. *Environmental Ethics* **23**, 155–168.

Azar, C., J. Holmberg and K. Lindgren (1996). Socio-ecological indicators for sustainability. *Ecological Economics* **18**, 89–112.

Bakkes, J. A., G. J. van den Born, J. C. Helder, R. J. Swart, C. W. Hope and J. D. E. Parker (1994). *An overview of environmental indicators: State of the Art and Perspectives*. UNEP/EATR.94-01; RIVM/402001001 Environmental Assessment Sub-Programme, Nairobi: UN Environmental Programme (UNEP).

Barbier, E. B. (1990). Alternative approaches to economic-environmental interactions. *Ecological Economics* **2**, 7–26.

(1994). Valuing environmental functions: tropical wetlands. *Land Economics* **70**, 155–173.

Barbier, E. B., J. C. Burgess and C. Folke (1994). *Paradise Lost? The Ecological Economics of Biodiversity*. London: Earthscan.

Barbier, E. B., M. Acreman and D. Knowler (1997). *Economic Valuation of Wetlands: A Guide for Policy Makers and Planners*. Gland, Switzerland: Ramsar Convention Bureau.

Barendregt, A. and J. W. Nieuwenhuis (1993). ICHORS, hydro-ecological relations by multi-dimensional modelling of observations. *Proceedings and Information*, No. 47. Delft, the Netherlands: CHO-TNO, pp. 11–30.

Barendregt, A., S. M. E. Stam and M. J. Wassen (1992). Restoration of fen ecosystems in the Vecht River plain: cost–benefit analysis of hydrological alternatives. *Hydrobiologia* **233**, 247–258. [Also in van Liere L. and R. D. Gulati (eds.) *Restoration and Recovery of Shallow Eutrophic Lake Ecosystems in the Netherlands*. Dordrecht: Kluwer.]

Barendregt, A., M. J. Wassen and J. T. de Smidt (1993). Hydro-ecological modelling in a polder landscape: a tool for wetland management. In C. Vos and P. Opdam (eds.) *Landscape Ecology of a Stressed Environment*. London: Chapman & Hall, pp. 79–99.

Batabyal, A. A. (1999). Aspects of the optimal management of cyclical ecological–economic systems. *Ecological Economics* **30**, 285–292.

Baumol, W. J. and W. E. Oates (1988). *The Theory of Environmental Policy*, 2nd edn. Cambridge: Cambridge University Press.

Beinat, E. (1995). Multiattribute Value Functions for Environmental Management. Ph.D. Thesis. Free University, Amsterdam.

Beltratti, A. (1996). *Models of Economic Growth with Environmental Assets*. Dordrecht: Kluwer.

Bennett, R. J. and R. J. Chorley (1978). *Environmental Systems*. London: Methuen.

Berkes, F. and C. Folke (eds.) (1998). *Linking Social and Ecological Systems: Management Practices and Social Mechanisms for Building Resilience*. Cambridge: Cambridge University Press.

Bertin, J. (1981). *Graphics and Graphic Information Processing*, Berlin: Walter de Gruyter.

Bibby, C. J. (1998). Selecting areas for conservation. In W. J. Sutherland (ed.) *Conservation Science and Action*. Oxford: Blackwell Science, pp. 176–201.

Biggs, D., J. B. Robinson and M. Walsh (1996). *Quest: A Quasi Understandable Ecosystem Scenario Tool*. Vancouver, BC: Sustainable Development Research Institute of Canada.

Bingham, G., R. Bishop, M. Brody, *et al.* (1995). Issues in ecosystem valuation: improving information for decision making. *Ecological Economics* **14**, 73–90.

Blamey, R. K. and M. S. Common (1999). Valuation and ethics in environmental economics. In J. C. J. M. van den Bergh (ed.) *Handbook of Environmental and Resource Economics*. Cheltenham, UK: Edward Elgar, pp. 809–823.

Bockstael, N., R. Costanza, I. Strand, W. Boynton, K. Bell and L. Waigner (1995). Ecological economic modeling and valuation of ecosystems. *Ecological Economics* **14**, 143–159.

Bornette, G., C. Amoros, H. Piegay, J. Tachet and T. Hein (1998). Ecological complexity of wetlands within a river landscape. *Biological Conservation* **85**, 35–45.

Bos, E. J. and J. C. J. M. van den Bergh (2002). A cost–benefit analysis of sustainable nature policy in the Dutch Vecht wetlands area. In R. Florax, P. Nijkamp and K. Willis (eds.) *Comparative Environmental Economic Assessment*. Cheltenham, UK: Edward Elgar, pp. 246–270.

Boulding, K. E. (1966). The economics of the coming spaceship earth. In H. Jarret (ed.) *Environmental Quality in a Growing Economy*. Baltimore, MD: Johns Hopkins University Press, pp. 3–14.

Braat, L. C. (1992). Sustainable Multiple Use of Forest Ecosystems: An Economic–Ecological Analysis for Forest Management in the Netherlands. Ph.D. Thesis, Vrije Universiteit, Amsterdam.

Braat, L. C. and W. F. J. van Lierop (eds.) (1987). *Economic–Ecological Modelling*. Amsterdam: North-Holland.

Brouwer, F. M. (1987). *Integrated Environmental Modelling: Design and Tools*. Dordrecht: Martinus Nijhoff.

Brouwer, R. (2000). Environmental value transfer: state of the art and future prospects. *Ecological Economics* **23**, 137–152.

Brouwer, R. and F. A. Spaninks (1999). The validity of environmental benefit transfer: further empirical testing, *Environmental and Resource Economics* **14**, 95–117.

Brouwer, R., I. H. Langford, I. J. Bateman and R. K. Turner (1999). A meta-analysis of wetland contingent valuation studies. *Regional Environmental Change* **1**, 47–57.

Bruce, J., H. Lee and E. F. Haites, for Working Group III (1995). *Climate Change 1995: Economic and Social Dimensions of Climate Change*. Cambridge: Cambridge University Press.

Button, K. J. and E. T. Verhoef (1998). *Road Pricing, Traffic Congestion and the Environment*. Cheltenham, UK: Edward Elgar.

Byström, O. (1998). The nitrogen abatement cost in wetlands, *Ecological Economics* **26**, 321–331.

Carpenter, S., W. Brock and P. Hanson (1999). Ecological and social dynamics in simple models of ecosystem management. *Conservation Ecology* **3**, 4.

Carson, R. T., W. M. Hanemann, R. J. Kopp, S. Presser and P. Ruud (1992). *A Contingent Valuation Study of Lost Passive Use Values Resulting from the Exxon Valdez Oil Spill*. Anchorage: Attorney General of the State of Alaska.

Cartaxana, P. and F. Catarino (1997). Allocation of nitrogen and carbon in an esturine salt marsh in Portugal. *Journal of Coastal Conservation* **3**, 27–34.

Castells Cabré, N. (1999). International Environmental Agreements: Institutional Innovation in European Transboundary Air Pollution Policies. Ph.D. Thesis, Vrije Universiteit, Amsterdam.

CBS (1996a). *Statistical Yearbook*. [Dutch.] 's Gravenhage, the Netherland: SDU Uitgeverij, Central Bureau of Statistics.

(1996b). *Vakantie van Nederlanders 1995*: The Haag: Central Bureau of Statistics.

(1996c). *Bevolking der gemeenten van Nederland op 1 januari 1996*, Den Haag.

(1997a). *Statistical Yearbook 1997*. Voorburg/Heerlen: Central Bureau of Statistics.

(1997b). *Dagrecreatie 1995/1996*: The Haag: Central Bureau of Statistics.

(1998). *Tabellen uit Land- en Tuinbouwcijfers 1998*. The Haag: Central Bureau of Statistics and Landbouw Economisch Instituut.

Chapin III, F. S., E. S. Zavaleta, V. T. Eviner, *et al.* (2000). Consequences of changing biodiversity. *Nature* **405**, 234–242.

Chow, Ven te (ed.) (1964). *Handbook of Applied Hydrology*. New York: McGraw Hill.

Ciriacy-Wantrup, S. V. (1952). *Resource Conservation: Economics and Politics*. Berkeley, CA: University of California Press.

Claessen F. A. M., F. Klijn, J. P. M. Witte and J. G. Nienhuis (1994). Ecosystems classification and hydro-ecological modelling for national water management. In F. Klijn (ed.) *Ecosystem Classification for Environmental Management*. Dordrecht: Kluwer Academic, pp. 199–222.

Clark, C. W. (1990). *Mathematical Bioeconomics: The Optimal Management of Renewable Resources*, 2nd edn. New York: John Wiley.

Cofino, W. (1994). Quality management of monitoring programmes. In M. Adriaanse, J. van der Kraats, P. G. Stoks and R. C. Ward. (eds.) *Proceedings of the International Workshop Monitoring Tailor-made*, Beekbergen, the Netherlands, pp. 178–187.

Common, M. (1988). Poverty and progress revisited. In D. Collard, D. W. Pearce and D. Ulph (eds.) *Economics, Growth and Sustainable Environments*. New York: St Martin's Press, pp. 15–39.

Common, M. and C. Perrings (1992). Towards an ecological economics of sustainability. *Ecological Economics* **6**, 7–34.

Cordell, H. K. and J. C. Bergstrom (1993). Comparison of recreation use values among alternative reservoir water level management scenarios. *Water Resources Research* **29**, 247–258.

Costanza, R., C. S. Farber, J. Maxwell (1989). Valuation and management of wetland ecosystems. *Ecological Economics* **1**, 335–361.

Costanza, R., B. O. Norton and B. Haskell (eds.) (1992). *Ecosystem Health: New Goals for Environmental Management*. Washington, DC: Island Press.

Costanza, R., L. Wainger, C. Folke and K.-G. Maler (1993). Modeling complex ecological economic systems: towards an evolutionary, dynamic understanding of people and nature. *BioScience* **43**, 545–555.

Costanza, R., R. d'Arge, R. de Groot *et al.* (1997a). The value of the world's ecosystem services and natural capital. *Nature* **387**, 253–260.

Costanza, R., C. Perrings and C. J. Cleveland (1997b). *The Development of Ecological Economics*. Cheltenham, UK: Edward Elgar.

Couclelis, H. (1985). Cellular worlds: a framework for modeling micro-macro dynamics. *Environment and Planning A* **17**, 585–596.

Couwenberg, J. and H. Joosten (1999). Pools as missing links: the role of nothing in the being of mires. In V. Standen, J. Tallis and R. Meade (eds.) *Patterned Mires and Mire Pools: Origin and Development; Flora and Fauna*. Durham, UK: British Ecological Society, pp. 87–102.

Creel, M. and J. Loomis (1992). Recreation value of water to wetlands in the San Joaquin Valley: linked multinomial logic and count data trip frequency models, *Water Resources Research* **28**, 2597–2606.

Crocker, T. D. (1995). Ecosystem functions, economics and the ability to function. In J. W. Milon and J. F. Shogren (eds.) *Integrating Economic and Ecological Indicators: Practical Methods for Environmental Policy Analysis*. Westport, CT: Praeger.

Crocker, T. D. and J. Tschirhart (1992). Ecosystems, externalities and economics. *Environmental and Resource Economics* **2**, 551–567.

Crowards, T. (1998). Safe minimum standards: costs and opportunities. *Ecological Economics* **25**, 291–302.

Cumberland, J. H. (1966). A regional inter-industry model for analysis of development objectives. *Papers of the Regional Science Association* **17**, 65–95.

Daatselaar C. H. G., D. W. de Hoop, H. Prins and B. W. Zaalmink (1990). *Bedrijfsvergelijkend Onderzoek naar de Benutting van Mineralen op Melkveebedrijven*. The Haag: Landbouw-Economisch Instituut.

Daly, H. E. (1968). On economics as a life science. *Journal of Political Economy* **76**, 392–406.

(1992). Allocation, distribution, and scale: towards an economics that is efficient, just and sustainable. *Ecological Economics* **6**, 185–193.

Daly, H. E. and W. Cobb (1989). *For the Common Good: Redirecting the Economy Toward Community, the Environment and a Sustainable Future*. Boston, MA: Beacon Press.

Deeley, D. M. and E. I. Paling (1998). In A. J. McComb and J. A. Davis (eds.) *Assessing the Ecological Health of Estuaries in Southwest Australia*. Adelaide: Wetlands for the Future, Gleneagles Publishing, pp. 257–271.

de Groot, A. W. M., K. H. S. van Buiren, I. W. D. Overtoom and M. Zijl (1998). *SEO Report 465: Natuurlijk Vermogen: Een Empirische Studie naar de Economische Waardering van Natuurgebieden in het Algemeen en de Oostvaardersplaseen in het Bijzonder.* [*Natural Capital: An Empirical Study in the Economic Valuation of Nature Areas in General and the Oostvaardersplassen in Particular.*] Amsterdam: Stichting voor Economisch Onderzoek (SEO).

de Mooij, R. A. (1999). The double dividend of an environmental tax reform. In J. C. J. M. van den Bergh (ed.) *Handbook of Environmental and Resource Economics*. Cheltenham, UK: Edward Elgar, pp. 293–306.

Desvousges, W. H., M. C. Naughton and G. R. Parsons (1992). Benefit transfer: conceptual problems in estimating water quality benefits using existing studies, *Water Resources Research* **28**, 675–683.

Dominico, P. A. and F. W. Schwartz (1990). *Physical and Chemical Hydrogeology*. New York: John Wiley.

Duchin, F. and G. M. Lange, in association with K. Thonstad and A. Idenburg (1994). *The Future of the Environment: Ecological Economics and Technical Change*. Oxford: Oxford University Press.

Dugan, P. J. (1990). *Wetland Conservation: A Review of Current Issues and Required Action*. Gland, Switzerland: IUCN.

Edward-Jones, G., B. Davies and S. Hussain (2000). *Ecological Economics, An Introduction*. Oxford: Blackwell Science.

Eggert, H. (1998). Bioeconomic analysis and management. *Environmental and Resource Economics* **11**, 399–411.

Ellery, W. N., T. S. McCarthy and J. M. Dangerfield (2000). Floristic diversity in the Okavango Delta, Botswana as an endogenous product of biological activity. In B. Gopal, W. J. Junk and J. A. Davis (eds.) *Biodiversity in Wetlands: Assessment, Function and Conservation*, Vol. 1. Leiden: Backhuys, pp. 195–226.

Engelen, G., R. White, I. Uljee and P. Drazan (1995). Using cellular automata for integrated modelling of socio-environmental systems. *Environmental Monitoring and Assessment* **34**, 203–214.

English Nature (1993). *Strategy for the 1990s*. Peterborough, UK: English Nature.

Esty, D. C. (1994). *Greening the GATT: Trade, Environment and the Future*. Washington, DC: Institute for International Economics.

Faber, M. and J. L. R. Proops (1990). *Evolution, Time, Production and the Environment*. Heidelberg: Springer-Verlag.

Fahrig, L. and G. Merriam (1985). Habitat patch connectivity and population survival. *Ecology* **66**, 1762–1768.

Folke, C. (1999). Ecological principles and environmental economic analysis. In J. C. J. M. van den Bergh (ed.) *Handbook of Environmental and Resource Economics*. Cheltenham, UK: Edward Elgar, pp. 895–911.

Freeman III, A. M. (1993). *The Measurement of Environmental and Resource Values: Theory and Methods*. Baltimore, MD: Resources for the Future.

Funtowicz, S. and S. Ravetz (1994). The worth of a songbird: ecological economics as a post-modern science. *Ecological Economics* **10**, 197–207.

Gaston, K. J. (1996). *Biodiversity: A Biology of Numbers and Differences.* Oxford: Blackwell Science.

(1998). Biodiversity. In W. J. Sutherland (ed.) *Conservation Science and Action,* Vols. 1–19. Oxford: Blackwell Science.

Georgescu-Roegen, N. (1971). *The Entropy Law and the Economic Process.* Cambridge, MA: Harvard University Press.

Ghilarov, A. (1996). What does 'biodiversity' mean: scientific problem or convenient myth? *Trends in Ecology and Evolution* **11**, 304–306.

Giaoutzi, M. and P. Nijkamp (1993). *Decision Support Models for Regional Sustainable Development.* Aldershot, UK: Avebury.

Gibson, C. C., E. Ostrom and T. K. Ahn (2000). The concept of scale and the human dimensions of global change: a survey. *Ecological Economics* **32**, 217–239.

Gilbert, A. J. and L. C. Braat (eds.) (1991). *Modelling for Population and Sustainable Development.* London: Routledge.

Gilbert, A. J. and J. F. Feenstra (1994). A sustainability indicator for the Dutch environmental policy theme 'Diffusion': cadmium accumulation in soil. *Ecological Economics* **9**, 253–265.

Gilbert, A. J. and R. Janssen (1998). Use of environmental functions to communicate the values of a mangrove ecosystem under different management regimes. *Ecological Economics* **25**, 323–346.

Gopal, B., W. J. Junk and J. A. Davis (eds.) (2000). *Biodiversity in Wetlands: Assessment, Function and Conservation,* Vol. 1. Leiden: Backhuys.

Gorham, E. (1991). Northern peatlands: role in the carbon cycle and probable responses to climatic warming. *Ecological Applications* **1**, 182–195.

Gosselink, J. G. and E. Maltby (1990). Wetland losses and gains. In M. Williams (ed.) *Wetlands, a Threatened Landscape.* Oxford: Basil Blackwell, pp. 296–322.

Gowdy, J. (1997). The value of biodiversity: markets, society and ecosystems. *Land Economics* **73**, 25–41.

(1999). Evolution, environment and economics. In J. C. J. M. van den Bergh (ed.) *Handbook of Environmental and Resource Economics.* Cheltenham, UK: Edward Elgar, pp. 965–980.

Gowdy, J. M. and A. Ferrer-i-Carbonell (1999). Toward consilience between biology and economics: the contribution of ecological economics. *Ecological Economics* **29**, 337–348.

Gradus, R. and S. Smulders (1993). The trade-off between environmental care and long-term growth: pollution in three proto-type growth models. *Journal of Economics* **58**, 25–52.

Gren, I.-G., C. Folke, K. Turner and I. Bateman (1994). Primary and secondary values of wetland ecosystems. *Environmental and Resource Economics* **4**, 55–74.

Grime, J. P. (1979). *Plant Strategies and Vegetation Processes.* Chichester, UK: John Wiley.

Grootjans, A. P., W. H. O. Ernst, P. J. Stuyfzand (1998). European dune slacks: strong interaction of biology, pedogenesis and hydrology. *Trends in Ecology and Evolution* **13**, 96–100.

Gunderson, L. H. and C. S. Holling (2001). *Panarchy: Understanding Transformations in Systems of Humans and Nature*. Washington DC: Island Press.

Gunderson, L. H., C. S. Holling and S. S. Light (eds.) (1995). *Barriers and Bridges to the Renewal of Ecosystems and Institutions*. New York: Columbia University Press.

Hafkamp, W. A. (1984). *Triple Layer Model: A National-Regional Economic-Environmental Model for the Netherlands*. Amsterdam: North-Holland.

Hall, C. A. S., G. Leclerc and C. Leon Perez (eds.) (2000). *Quantifying Sustainable Development: The Future of Tropical Economies*. New York: Academic Press.

Hanley, N. (1999). Cost–benefit analysis of environmental policy and management. In J. C. J. M. van den Bergh (ed.) *Handbook of Environmental and Resource Economics*, Cheltenham, UK: Edward Elgar, pp. 824–836

Hanley, N. and C. L. Spash (1993). *Cost–Benefit Analysis and the Environment*. Aldershot, UK: Edward Elgar.

Hannon, B. (1973). The structure of ecosystems. *Journal of Theoretical Biology* **41**, 535–546.

(1976). Marginal product pricing in the ecosystem. *Journal of Theoretical Biology* **56**, 253–267.

(1986). Ecosystem control theory. *Journal of Theoretical Biology* **121**, 417–437.

(1991). Accounting in ecological systems. In R. Costanza (ed.) *Ecological Economics: The Science and Management of Sustainability*. New York: Columbia University Press.

Hanski, I. (1998). Metapopulation dynamics. *Nature* **396**, 41–49.

Hanski, I. and M. E. Gilpin (eds.) (1997). *Metapopulation Biology, Ecology, Genetics, and Evolution*. San Diego, CA: Academic Press.

Harper, J. L. and D. L. Hawksworth (1995). *Biodiversity, measurement and estimation*. London: Chapman & Hall.

Hausman, J. A. (ed.) (1993). *Contingent Valuation: A Critical Assessment*. Amsterdam: North-Holland.

Heij, G. J. and G. Bannink (1995). *River Basin Fresh Water Resources Assessment Methodology*. Bilthoven, the Netherlands: National Institute of Public Health and the Environment (RIVM).

Heij, G. J., M. J. H. Pastoors and H. Snelting (1985). De berekening van stroombanen en intrekgebieden in het gebied van de Utrechtse Heuvelrug. *H20*, **18**, 182–186.

Heijnens, F. J. (1988). *Technische Toelichting Berekeningen Grondwater-Beschermings-Gebieden*. Leidschendam, the Netherlands: RIVM.

Helmer, W., P. Vellinga, G. Luitjens, H. Goosen, E. Ruijgrok and W. Overmars (1996). *Growing with the Sea: Towards a Resilient Coast*. Zeist: World Wildlife Fund.

Herendeen, R. A. (1999). EMERGY, value and economics. In J. C. J. M. van den Bergh (ed.) *Handbook of Environmental and Resource Economics*. Cheltenham, UK: Edward Elgar, pp. 954–964.

Hill, M. C. (1990). *Preconditionated Conjugate Gradient 2 (PCG2), A Computer Program for Solving Groundwater Flow Equations. Water-Resources Investigation Report* 90-4048. Reston, VA: US Geological Survey.

Holling, C. S. (1973). Resilience and stability of ecological systems. *Annual Review of Ecological Systems* **4**, 1–24.

(1978). *Adaptive Environmental Assessment and Management.* New York: John Wiley.

(1986). The resilience of terrestrial ecosystems: local surprise and global change. In W. C. Clark and R. E. Munn (eds.) *Sustainable Development of the Biosphere.* Cambridge: Cambridge University Press.

(1998). Two cultures of ecology. *Conservation Ecology* **2**, 4. http://www.consecol.org.

Holling, C. S., D. W. Schindler, B. Walker and J. Roughgarden (1995). Biodiversity in the functioning of ecosystems: an ecological primer and synthesis. In C. Perrings, K.-G. Mäler, C. Folke, C. S. Holling and B.-O. Jansson (eds.) *Biodiversity Loss: Ecological and Economic Issues.* Cambridge: Cambridge University Press.

Hosper, H. (1997). *Clearing Lakes, An Ecosystem Approach to the Restoration and Management of Shallow Lakes in the Netherlands.* Ph.D. Thesis. Agricultural University of Wageningen, the Netherlands.

Huisman, J., and F. J. Weissing (1999). Biodiversity of plankton by species oscillations and chaos. *Nature* **402**, 407–410.

Hutcheson, K. (1970). A test for comparing diversities based on the Shannon formula. *Journal of Theoretical Biology* **29**, 151–154.

Isard, W. (1969). Some notes on the linkage of ecologic and economic systems. *Papers of the Regional Science Association* **22**, 85–96.

(1972). *Ecologic–Economic Analysis for Regional Development.* New York: The Free Press.

IUCN (1993). *IUCN Red Lists of Threatened Animals 1994* (B. Groonbridge (ed.)). Gland, Switzerland: International Union for the Conservation of Nature and Natural Resources.

IUCN, UNEP and WWF (1980). *World Conservation Strategy: Living resource conservation for sustainable development.* Morges, Switzerland: IUCN.

IWACO (1992). *Grondwaterbeheer Midden Nederland. Modellering Watersysteem.* Rotterdam: IWACO.

(1994). *Hydrologisch onderzoek Horstermeerpolder.* Rotterdam: IWACO.

Janssen, M. A. (1998a). Use of complex adaptive systems for modeling global change. *Ecosystems* **1**, 457–463.

(1998b). *Modelling Global Change: The Art of Integrated Assessment.* Cheltenham, UK: Edward Elgar.

Janssen, M. A. and H. J. M. de Vries (1998). The battle of perspectives: a multi-agents model with adaptive responses to climate change. *Ecological Economics* **26**, 43–65.

Janssen, M. A., B. H. Walker, J. Langridge and N. Abel (1999). *An Integrated Model of Rangelands as a Complex Adaptive System.* [Working paper.] Canberra: CSIRO Wildlife and Ecology.

Janssen, R. (1992). *Multiobjective Decision Support for Environmental Management.* Dordrecht: Kluwer.

Janssen, R. and G. Munda (1999). Multi-criteria methods for quantitative, qualitative and fuzzy evaluation problems. In J. C. J. M. van den Bergh (ed.) *Handbook of Environmental and Resource Economics*. Cheltenham, UK: Edward Elgar, pp. 837–852.

Janssen, R. and J. E. Padilla (eds.) (1997). *Mangroves or Fishponds? Valuation and Evaluation of Management Alternatives for the Pagbilao Mangrove Forest (Final Report)*. Pasig City, Philippines: Resources, Environment and Economics Centre for Studies Inc.: Amsterdam: Institute for Environmental Studies, Vrije Universiteit.

Janssen, R. and M. van Herwijnen (1994). *DEFINITE, a System to Support Decisions on a Finite Set of Alternatives*. Dordrecht: Kluwer.

Jansson, A. M. and J. Zucchetto (1978). *Ecological Bulletins No. 28: Energy, Economic and Ecological Relationships for Gotland, Sweden: A Regional Systems Study*. Stockholm: Swedish Natural Science Research Council.

Jørgenson, S. E. (1992). *Integration of Ecosystem Theories: A Pattern*. Dordrecht: Kluwer.

Kamien, M. I. and N. L. Schwartz (1982). The role of common property resources in optimal planning models with exhaustible resources. In V. K. Smith and J. V. Krutilla (eds.) *Explorations in Natural Resource Economics*. Baltimore, MD: Johns Hopkins University Press.

Karr, J. R. (1981). Assessment of biotic integrity using fish communities. *Fisheries* **6**, 21–27.

(1991). Biological integrity: a long-neglected aspect of water resource management. *Ecological Applications* **1**, 66–84.

Klein, R. J. T. and I. J. Bateman (1998). The recreation value of Cley marshes nature reserve: an argument against managed retreat. *Water and Environmental Management* **12**, 280–285.

Koerselman, W. and B. Beltman (1988). Evapotranspiration from fens in relation to Penman's potentional free water evaporation (EÔ) and pan evaporation. *Aquatic Botany* **31**, 307–320.

Können, G. P. (1999). *De Toestand van het Klimaat in Nederland 1999. [The State of the Climate in the Netherlands 1999.]* De Bilt, the Netherlands: KNMI.

Kraak, M. J. and F. J. Ormeling (1996). *Cartography: Visualization of Spatial Data*, Harlow, UK: Addison Wesley Longman.

Krebs, C. J. (1972). *Ecology, The Experimental Analysis of Distribution and Abundance*. New York: Harper & Row.

Kuik, O. and H. Verbruggen (eds.) (1991). *In Search of Indicators for Sustainable Development*. Dordrecht: Kluwer.

Lakshmanan, T. R. and R. Bolton (1986). Regional energy an environmental analysis. In P. Nijkamp (ed.) *Handbook of Regional and Urban Economics*, Vol. 1. Amsterdam: North-Holland.

Lee, K. N. (1993). *Compass and Gyroscope: Integrating Science and Politics for the Environment*. Washington, DC: Island Press.

LEI (1998). *Bedrijfsuitkomsten in de Landbouw*. The Haag: Landbouw Economisch Instituut.

Leontief, W., A. Carter and P. Petri (1977). *Future of the World Economy*. New York: Oxford University Press.

Levin, S. A., S. Barrett, S. Aniyat *et al.* (1998). Resilience in natural and socio-economic systems. *Environment and Development Economics* **3**, 222–235.

Likens, G. E., F. H. Bormann, R. S. Pierce, J. S. Eaton and N. M. Johnson (1977). *Biogeochemistry of a Forested Ecosystem*. New York: Springer Verlag.

Liverman, D. M., M. E. Hanson, B. J. Brown and R. W. Meredith Jr (1988). Global sustainability: towards measurement, *Environmental Management* **12**, 133–143.

LNV (1990). *Natuurbeleidsplan*. [*Nature Policy Plan*.] The Haag: Ministerie van Landbouw, Natuur en Visserij (Ministry of Agriculture, Nature and Fisheries).

(1993). *Structuurschema Groene Ruimte*. [*Plan for Green Space*.] The Haag: Ministerie van Landbouw, Natuur en Visserij (Ministry of Agriculture, Nature and Fisheries).

(1996). *Structuurschema Groene Ruimte*. [*Report on Green Space*.] The Haag: Ministerie van Landbouw, Natuur en Visserij (Ministry of Agriculture, Nature and Fisheries).

Loomis, J. B. (1987). Balancing public trust resources of Mono Lake and Los Angeles' Water Right: an economic approach. *Water Resources Research*, **23**, 1449–1456.

Lorenz, C. M. (1999). Indicators for Sustainable River Management. Ph.D. Thesis, Vrije Universiteit, Amsterdam.

Lorenz, C. M., A. J. Gilbert and P. Vellinga (2001a). Sustainable management of transboundary river basins: a line of reasoning. *Regional Environmental Change* **2**, 38–53.

Lorenz, C. M., A. J. Gilbert and W. P. Cofino (2001b). Indicators for transboundary river management. *Environmental Management* **28**, 115–129.

MacArthur, R. H. and E. O. Wilson (1967). *Island Biogeography*. Princeton, NJ: Princeton University Press.

MacEachren, A. M. (1994). *Some Truth with Maps: A Primer on Symbolization and Design*. Washington, DC: Association of American Geographers.

Mageau, M. T., R. Costanza and R. E. Ulanowicz (1998). Quantifying the trends expected in developing ecosystems. *Ecological Modelling* **112**, 1–22.

Maltby, E. (1986). *Waterlogged Wealth: Why Waste the World's Wet Places?* London: Earthscan.

Maltby, E., D. V. Hogan and R. J. McInnes (eds.) (1996). *Functional Analysis of European Wetland Ecosystems. Final Report – Phase One*. [EC DGXII STEP Project CT90-0084.] London: Wetland Ecosystems Research Group, University of London.

Manning, J. C. (1997). *Applied Principles of Hydrology*, 3rd edn. Upper Saddle River, NJ: Prentice Hall.

Martinez-Alier, J. and M. O'Connor (1999). Distribution issues: an overview. In: J. C. J. M. van den Bergh (ed.) *Handbook of Environmental and Resource Economics*. Cheltenham, UK: Edward Elgar, pp. 380–392.

Martinez-Alier, J., G. Munda and J. O'Neill (1998). Weak comparability of values as a foundation for ecological economics. *Ecological Economics* **26**, 277–286.

Maynard Smith, J. (1974). *Models in Ecology.* Cambridge: Cambridge University Press.

McConnell, K. E. (1992). Model building and judgement: implications for benefit transfers with travel cost models, *Water Resources Research* **28**, 695–700.

McDonald, M. G. and A. W. Harbaugh (1984). *A Modular Three-dimensional Finite-difference Groundwater Flow Model.* Richmond, VA: US Geological Survey (USGS).

McDonald, M. G., A. W. Harbaugh, R. O. Brennon and D. J. Ackerman (1991). *A Method for Converting No-flow Cells to Variable Head Cells for the U.S. Ecological Survey Modular Finite-difference Groundwater Flow Model.* Richmond, VA: US Geological Survey (USGS).

Meadows, D. H., D. L. Meadows, J. Randers and W. W. Behrens III (1972). *The Limits to Growth.* New York: Universe Books.

Meadows, D. H., J. Richardson and G. Bruckmann (1982). *Groping in the Dark: The First Decade of Global Modeling.* New York: John Wiley.

Meadows, D. H., D. L. Meadows and J. Randers (1992). *Beyond the Limits: Confronting Global Collapse; Envisioning a Sustainable Future.* Post Mills: Chelsea Green.

Millennium Institute (1996). *Documentation for Threshold 21: National Sustainable Development Model.* Arlington, DC: Millenium Institute.

Mitsch, W. J. and J. G. Gosselink (1993). *Wetlands.* New York: Van Nostrand Reinhold.

Monmonier, M. S. (1991). *How to Lie with Maps.* Chicago, IL: The University of Chicago Press.

Moore, P. D. and D. J. Bellamy (1974). *Peatlands.* London: Elek Science.

Moreira, F., P. G. Ferreira, F. C. Rego and S. Bunting (2001). Landscape changes and breeding bird assemblages in northwestern Portugal: the role of fire. *Landscape Ecology* **16**, 175–187.

Moss, B. (1988). *Ecology of Fresh Waters: Man and Medium*, 2nd edn. Oxford: Blackwell Scientific.

Munda, G. (1995). *Multicriteria Evaluation in a Fuzzy Environment.* Heidelberg: Physica-Verlag.

(1997). Multicriteria evaluation as a multidimensional approach to welfare measurement. In J. C. J. M. van den Bergh and J. van der Straaten (eds.) *Economy and Ecosystems in Change: Analytical and Historical Approaches.* Cheltenham, UK: Edward Elgar, pp. 96–115.

Munda, G., P. Nijkamp and P. Rietveld (1994). Qualitative multi-criteria evaluation for environmental management. *Ecological Economics* **10**, 97–112.

Munro, A. (1997). Economics and biological evolution. *Environmental and Resource Economics* **9**, 429–449.

Muradian, R. (2001). Ecological thresholds: a survey. *Ecological Economics* **38**, 7–24.

National Wetland Working Group (1988). *Ecological Land Classification Series*, No. 24: *Wetlands of Canada.* Montreal: Sustainable Development Branch, Environment Canada and Polyscience Publications.

Neher, P. A. (1990). *Natural Resource Economics: Conservation and Exploitation.* New York: Cambridge University Press.

Nelder, J. A. and R. W. M. Wetherburn (1974). Generalized linear models. *Journal of the Royal Statistical Society of London A* **135**, 370–384.

Nijkamp, P. (1979a). *Theory and Application of Environmental Economics*. Amsterdam: North-Holland.

(1979b). *Environmental Policy Analysis*. Chichester, UK: John Wiley.

Nijkamp, P., P. Rietveld and F. Snickars (1986). Regional and multiregional economic models: a survey. In P. Nijkamp (ed.) *Handbook of Regional and Urban Economics*, Vol. 1. Amsterdam: North-Holland.

Nijkamp, P., P. Rietveld and H. Voogd (1990). *Multicriteria Evaluation in Physical Planning*. Amsterdam: North-Holland.

Nordhaus, W. D. (1991). To slow or not to slow: the economics of the greenhouse effect. *Economic Journal* **101**, 920–937.

(1994). *Managing the Global Commons: The Economics of Climate Change*. Cambridge, MA: MIT Press.

Norgaard, R. B. (1994). *Development Betrayed: The End of Progress and a Coevolutionary Revisioning of the Future*. London: Routledge.

Nunes, P. A. L. D., and J. C. J. M. van den Bergh (2001). Monetary valuation of biodiversity: sense or nonsense? *Ecological Economics* **39**, 203–222.

Nunes, P. A. L. D., J. C. J. M. van den Bergh and P. Nijkamp (2002). *The Ecological Economics of Biodiversity: Methods and Applications*. Cheltenham, UK: Edward Elgar.

Odum, E. P. (1975). *Ecology: The Link Between the Natural and Social Sciences*. London: Holt Rinehart and Winston.

(1983). *Systems Ecology: An Introduction*. New York: John Wiley.

(1987). Models for national, international and global systems policy. In L. C. Braat and W. F. J. van Lierop (eds.) *Economic–Ecological Modelling*. Amsterdam: North-Holland, pp. 203–251.

OECD (1994). *Environmental Indicators: OECD Core Set*. Paris: Organization for Economic Cooperation and Development.

(1995). *Environmental Data: Compendium 1995*. Paris: Organization for Economic Cooperation and Development.

Opschoor, J. B. and R. Weterings (eds.) (1994). Environmental utilisation space. *Milieu (The Netherlands' Journal of Environmental Science Special Issue)* **9** [In English].

O'Riordan, T. (ed.) (1997). *Ecotaxation*. London: Earthscan.

Parry, M. and T. Carter (1998). *Climate Impact and Adaptation Assessment*. London: Earthscan.

Parsons, G. R. and M. J. Kealy (1994). Benefits transfer in a random utility model of recreation, *Water Resource Research*, **30**, 2477–2484.

Patten, B. C. (ed.) (1971). *Systems Analysis and Simulation in Ecology*, Vols. I–IV. London: Academic Press.

Patterson, E. (1992). GATT and the environment: rules changes to minimize adverse trade and environmental effects, *Journal of World Trade Law* **26**, 100–110.

Pearce, D. (1998). Auditing the earth. *Environment* **2**, 23–28.

Pearce, D. W. and C. A. Perrings (1995). Biodiversity conservation and economic development: local and global dimensions. In C. A. Perrings, K.-G. Mähler,

C. Floke, C. S. Holling and B.-O. Jansson (eds.) *Biodiversity Conservation: Problems and Policies*. Dordrecht: Kluwer.

Pearson, M. (1992). *Recreation and Environmental Valuation of Rutland Water*. Norwich, UK: Centre for Social and Economic Research on the Global Environment (CSERGE).

Perrings, C. (1987). *Economy and Environment*. New York: Cambridge University Press.

(1998). Resilience in the dynamics of economic-environment systems. *Environmental and Resource Economics* **11**, 503–520.

Perrings, C. and D. W. Pearce (1994). Threshold effects and incentives for the conservation of biodiversity. *Environmental and Resource Economics* **4**, 13–28.

Perrings, C. and D. I. Stern (2000). Modelling loss of resilience in agroecosystems: rangeland in Botswana. *Environmental and Resource Economics* **16**, 185–210.

Perrings, C. and B. Walker (1997). Biodiversity, resilience and the control of ecological–economic systems: the case of fire-driven rangelands. *Ecological Economics* **22**, 73–83.

Perrings, C., K.-G. Mäler, C. Folke, C. S. Holling and B.-O. Jansson (eds.) (1995a). *Biodiversity Conservation: Problems and Policies*. Dordrecht: Kluwer.

(1995b). *Biodiversity Loss*. Cambridge: Cambridge University Press.

Pimm, S. L. (1984). The complexity and stability of ecosystems. *Nature* **307**, 321–326.

Pimm, S. L. and J. H. Lawton (1998). Planning for biodiversity. *Science* **279**, 2068–2069.

Plaziat, J-C., C. Cavagnetto, J-C. Koeniguer and F. Latzer (2001). History and biogeography of the mangrove ecosystem, based on a critical reasssessment of the paleontological record. *Wetlands Ecology and Management* **9**, 161–179.

Pons, L. J. (1992). Holocene peat formation in the lower parts of the Netherlands. In J. T. A. Verhoeven (ed.) *Fens and Bogs in the Netherlands: Vegetation, History, Nutrient Dynamics and Conservation*. Dordrecht: Kluwer Academic, pp. 7–80.

Porter, R. (1982). The new approach to wilderness appraisal through cost–benefit analysis. *Journal of Environmental Economics and Management* **11**, 59–80.

Price, C. (1993). *Time, Discounting and Value*. Oxford: Blackwell.

Provincie Noord-Holland (1995). *Gebiedsperspectief voor de Vechtstreek*. [*Regional Plan for the Vecht Area*.] Haarlem: Beleidsnota, Provincie Noord-Holland.

(1997). *Kerncijfers Land en Tuinbouw Noord-Holland 1997*. [*Core Data Agriculture and Horticulture North-Holland 1997*.] Haarlem: Provincie Noord-Holland. [Core statistics concerning agriculture in North-Holland.]

Prudic, D. E. (1989). *Documentation of a Computer Program to Simulate Stream Aquifer Relations using a Modular, Finite-difference Groundwater Flow Model*. Carson City, NV: US Geological Survey (USGS).

Rapport, D. J., R. Costanza and A. J. McMichael (1998). Assessing ecosystem health. *Trends in Ecology and Evolution* **13**, 397–402.

Rapport, D. J., A. McMichael and R. Costanza (1999). Reply from D. J. Rapport, A. J. McMichael and R. Costanza. *Trends in Ecology and Evolution* **14**, 69–70.

Rawls, J. (1972). *A Theory of Justice*. Cambridge, MA: Harvard University Press.

Rietkerk, M., P. Ketner, J. Burger, B. Hoorens, and H. Olff (2000). Multiscale soil and vegetation patchiness along a gradient of herbivore impact in a semi-arid grazing system in West Africa. *Plant Ecology* **148**, 207–224.

Rietveld, P. (1990). Multidimensional inequality comparisons: on aggravation and mitigation of inequalities. *Economic Letters* **32**, 187–192.

(1991). A note on interregional versus intraregional inequality. *Regional Science and Urban Economics* **21**, 627–637.

Rose, A. and S. Kverndokk (1999). Equity in environmental policy: an application to global warming. In J. C. J. M.. van den Bergh (ed.) *Handbook of Environmental and Resource Economics*. Cheltenham, UK: Edward Elgar, pp. 352–379.

Rotmans, J. (1997). Indicators for sustainable development. In J. Rotmans and H. J. M. de Vries (eds.) *Perspectives on Global Change: The TARGETS Approach*. Cambridge: Cambridge University Press.

Rotmans, J. and B. de Vries (1997). *Perspectives on Global Change: The TARGETS Approach*. Cambridge: Cambridge University Press.

Rotmans, J. and P. Vellinga (eds.) (1998). Challenges and opportunities for integrated environmental assessment. *Environmental Modeling and Assessment*, **3**. *Special Issue*.

Roughgarden, R., R. M. May and S. A. Levin (eds.) (1989). *Perspectives in Ecological Theory*. Princeton, NJ: Princeton University Press.

Ruijgrok, E. C. M. (1998). *Valuation of Nature in Coastal Zones*, Part IV: *Perception Value*, Gouda: Land Water Milieu Informatietechnologie.

Russell, C. S. (1995). Old lessons and new contexts in economic–ecological modeling. In J. W. Milon and J. F. Shogren (eds.) *Integrating Economic and Ecological Indicators: Practical Methods for Environmental Policy Analysis*. London: Praeger, pp. 9–25.

Ruth, M. (1993). *Integrating Economics, Ecology and Thermodynamics*. Dordrecht: Kluwer.

Sargent, T. J. and N. Wallace (1976). Rational expectations and the theory of economic policy. *Journal of Monetary Economics* **2**, 169–183.

Schaeffer, D. J., E. E. Herrinks and H. W. Kerster (1988). Ecosystem health: 1 Measuring ecosystem health. *Environmental Management* **12**, 445–455.

Scheffer, F. and P. Schachtschabel (1984). *Lehrbuch der Bodenkunde*, 11th edn. Stuttgart: Ferd. Enke-Verlag.

Scheffer, M., S. H. Hosper, M. L. Meijer and B. Moss (1993). Alternative equilibria in shallow lakes. *Trends in Ecology and Evolution* **8**, 275–279.

Scholten, H. J. and J. C. H. Stillwell (1990). *Geographical Information Systems for Urban and Regional Planning*. Dordrecht: Kluwer.

Schot, P. P. (1989). *Grondwatersystemen en Grondwaterkwaliteit in het Gooi en Randgebieden*. [In Dutch.] Utrecht: Interfaculteit Vakgroep Milieukunde.

(1991). *Solute Transport by Groundwaterflow to Wetland Ecosystems*. Ph.D. Thesis, Universitat, Utrecht.

(1999). Wetlands. In B. Nath, L. Hens, P. Compton and D. Devuyst (eds.) *Managing the Ecosystem: Environmental Management in Practice*, Vol. 3, London: Routledge, pp. 62–85.

Schot, P. P. and A. Molenaar (1992). Regional changes in groundwater flow patterns and effects on groundwater composition. *Journal of Hydrology* **130**, 151–170.

Schot, P. P. and J.van der Wal (1992). Human impact on regional groundwater composition through intervention in natural flow patterns and changes in land use. *Journal of Hydrology* **134**, 297–313.

Shackley, S. (1997). Trust in models? The mediating and transformative role of computer models in environmental discourse. In M. Redclift and G. Woodgate (eds.) *The International Handbook of Environmental Sociology*. Cheltenham, UK: Edward Elgar, pp. 237–260.

Shogren, J. F. and T. M. Hurley (1999). Experiments in environmental economics. In J. C. J. M. van den Bergh (ed.) *Handbook of Environmental and Resource Economics*, Cheltenham, UK: Edward Elgar, pp. 1180–1190.

Shogren, J. F. and C. Nowell (1992). Economics and ecology: a comparison of experimental methodologies and philosophies. *Ecological Economics* **5**, 101–126.

Siebert, H. (1982). Nature as a life support system: renewable resources and environmental disruption. *Journal of Economics* **42**, 133–142.

Simberloff, D. (1998). Small and declining populations. In W. J. Sutherland (ed.) *Conservation Science and Action*, Oxford: Blackwell Science, pp. 116–134.

Simpson, E. H. (1949). Measurement of diversity. *Nature* **163**, 688.

Spash, C. and N. Hanley (1995). Preferences, information and biodiversity preservation. *Ecological Economics* **12**, 191–208.

Speelman, H. and H. Houtman (1979). *Grondwaterkaart van Nederland. [Groundwater map of the Netherlands.]* Zandvoort (24), Amsterdam (25 west, 25 oost).

Spellerberg, I. F. (1996). *Conservation Biology*. Harlow, UK: Longman.

Steedman, R. J. (1994). Ecosystem health as management goal. *Journal of the North American Benthological Society* **13**, 605–610.

Stern, D. (1997). Limits to substitution and irreversibility in production and consumption: a neoclassical interpretation of ecological economics. *Ecological Economics* **22**, 197–215.

Stowa (1994). *Stowa Report*, No. 94-15: *Verwijdering van Fosfaat uit Oppervlaktewater: Evaluatie van in Nederland Toegepaste Systemen.* [In Dutch; *Removal of Phosphate from Surface Water: Evaluation of Systems Applied in the Netherlands.*] Utrecht: Stichting Toegepast Onderzoek Waterbeheer.

Succow, M. (1988). *Landschaftsokologische Moorkunde*. Berlin: Gebrüder Borntraeger.

Succow, M. and L. Jeschke (1986). *Moore in der Landschaft: Entstehung, Haushalt, Lebewelt, Verbreitung, Nutzung und Erhaltung der Moore.* Leipzig: Urania-Verlag.

Swallow, S. K. (1994). Renewable and nonrenewable resource theory applied to coastal agriculture, forest, wetland and fishery linkages. *Marine Resource Economics* **9**, 291–310.

Swanson, T. (1994). The economics of extinction revisited and revised: a generalised framework for the analysis of the problem of endangered species and biodiversity losses. *Oxford Economic Papers* **46** (Supplementary Issue), 800–821.

Swart, R. J. and J. A. Bakkes (eds.) (1995). *Scanning the Global Environment: A Framework and Methodology for Integrated Environmental Reporting and Assessment.* [UNEP/EATR.95-01; RIVM 402001002 Environmental Assessment Sub-Programme. UNEP] Nairobi: UN Environment Programme (UNEP).

Tahvonen, O. and J. Kuuluvainen (1993). Economic growth, pollution and renewable resources. *Journal of Environmental Economics and Management* **24**, 101–118.

Ten Brink, B. J. E., S. H. Hosper and F. Colijn (1991). A quantitative method for description and assessment of ecosystems: the AMOEBE approach. *Marine Pollution Bulletin* **23**, 265–270.

Tietenberg, T. (1996). *Environmental and Natural Resource Economics*, 4th edn. New York: Harper Collins.

Tockner, K., C. Baumgartner, F. Schiemer and J. V. Ward (2000). Biodiversity of a Danubian floodplain: structural, functional and compositional aspects. In B. Gopal, W. J. Junk and J. A. Davis (eds.) *Biodiversity in Wetlands: Assessment, Function and Conservation*, Vol. 1. Leiden: Backhuys, pp. 141–159.

Tol, R. (1998). Economic aspects of global environmental models. In J. C. J. M. van den Bergh and M. W. Hofkes (eds.) *Theory and Implementation of Economic Models for Sustainable Development*. Dordrecht: Kluwer, pp. 277–286.

Toman, M. A., J. Pezzey and J. Krautkraemer (1995). Neoclassical economic growth theory and 'sustainability'. In D. Bromley (ed.) *Handbook of Environmental Economics*. Oxford: Blackwell, pp. 139–165.

Turner, M. G. (1998). Landscape ecology. In S. I. Dodson, T. Allen, S. Carpenter *et al.* (eds.) *Ecology*. Oxford: Oxford University Press, pp. 77–122.

Turner, R. K. (1988). Wetland conservation: economics and ethics. In D. Collard, D. W. Pearce and D. Ulph (eds.) *Economics, Growth and Sustainable Environments*. London: Macmillan, pp. 121–159.

Turner, R. K., and T. Jones (eds.) (1991). *Wetlands, Market and Intervention Failures*. London: Earthscan.

Turner, R. K., C. Perrings and C. Folke (1997). Ecological economics: paradigm or perspective. In J. C. J. M. van den Bergh and J. van der Straaten (eds.) *Economy and Ecosystems in Change: Analytical and Historical Approaches*. Cheltenham, UK: Edward Elgar, pp. 25–49.

Turner, R. K., W. N. Adger and R. Brouwer (1998a). Ecosystem services value, research needs, and policy relevance: a commentary. Special section: forum on valuation of ecosystem services. *Ecological Economics* **25**, 61–65.

Turner, R. K., J. C. J. M. van den Bergh, A. Barendregt and E. Maltby (1998b). Ecological–economic analysis of wetlands: science and social science integration. In T. Söderqvist (ed.) *Beijer Occasional Paper Series, Wetlands: Landscape and Institutional Perspectives*. Stockholm, Sweden: Beijer International Institute of Ecological Economics, pp. 327–350. [Also as *Discussion Paper TI98-050/3*. Amsterdam: Tinbergen Institute.]

Turner, R. K., K. Button and P. Nijkamp (eds.) (1999). *Ecosystems and Nature: Economics, Science and Policy*. Cheltenham, UK: Edward Elgar.

Turner, R. K., J. C. J. M. van den Bergh, T. Söderqvist *et al.* (2000). Ecological–economic analysis of wetlands: scientific integration for management and policy. *Ecological Economics* **35**, 7–23.

V&W (1989). *Derde Nota Waterhuishouding. [Third Report on Water Management.]* The Haag: Ministerie van Verkeer en Waterstaat (Ministry of Transport).

van den Bergh, J. C. J. M., (1993). A framework for modelling economy–environment–development relationships based on dynamic carrying capacity and sustainable development feedback. *Environmental and Resource Economics* **3**, 395–412.

(1996). *Ecological Economics and Sustainable Development: Theory, Methods and Applications*. Cheltenham, UK: Edward Elgar.

(ed.) (1999). *Handbook of Environmental and Resource Economics*. Cheltenham, UK: Edward Elgar.

(2001). Ecological economics: themes, approaches, and differences with environmental economics. *Regional Environmental Change* **2**, 13–23.

van den Bergh, J. C. J. M. and R. A. de Mooij (1999). An assessment of the growth debate. In J. C. J. M. van den Bergh (ed.) *Handbook of Environmental and Resource Economics*. Cheltenham, UK: Edward Elgar, pp. 643–655.

van den Bergh, J. C. J. M. and J. M. Gowdy (2000). Evolutionary theories in environmental and resource economics: approaches and applications. *Environmental and Resource Economics* **17**, 37–57.

(2003). The microfoundations of macroeconomics: an evolutionary perspective. *Cambridge Journal of Economics* **27**, 65–84.

van den Bergh, J. C. J. M. and M. W. Hofkes (eds.) (1998). *Theory and Implementation of Economic Models for Sustainable Development*. Dordrecht: Kluwer.

van den Bergh, J. C. J. M. and P. Nijkamp (1991). Aggregate economic–ecological models for sustainable development. *Environment and Planning A* **23**, 187–206.

(1994a). Dynamic macro modelling and materials balance: economic–environmental integration for sustainable development. *Economic Modelling* **11**, 283–307.

(1994b). An integrated model for economic development and natural environment: an application to the Greek Sporades Islands. *Annals of Operations Research* **54**, 143–174.

van den Bergh, J. C. J. M. and H. Verbruggen (1999). Spatial sustainability, trade and indicators: an evaluation of the 'ecological footprint'. *Ecological Economics* **29**, 63–74.

van den Bergh, J. C. J. M., K. J. Button, P. Nijkamp and G. J. Pepping (1997). *Meta-analysis in Environmental Economics*. Dordrecht: Kluwer.

van den Bergh, J. C. J. M., A. Ferrer-i-Carbonell and G. Munda (2000). Alternative models of individual behaviour and implications for environmental policy. *Ecological Economics* **32**, 43–61.

van den Bergh, J. C. J. M., A. Barendregt, A. Gilbert, M. van Herwijnen, P. van Morssen, P. Kandelaars and C. Lorenz (2001). Spatial economic–hydroecological modelling and evaluation of land use impacts in the Vecht wetlands area. *Environmental Modeling and Assessment* **6**, 87–100.

van der Gun, J. A. M. (1978). *Grondwaterkaart van Nederland. [Groundwater map of the Netherlands.]* Utrecht (31 oost, 32 west, 38 oost, 39 west).

van der Maarel, E. (1980). *Succession. Advances in Vegetation Science 3*. The Hague: Junk.

van der Ploeg, S. W. F. (1990). Multiple Use of Natural Resources. Ph.D. Thesis, Vrije Universiteit, Amsterdam.

van der Valk, A. (ed.) (1989). *Northern Prairie Wetlands.* Ames, IO: Iowa State University Press.

van Dijk, P. M. and F. J. P. M. Kwaad (1998). Modelling suspended sediment supply to the river Rhine drainage network; a methodological study. In W. Summer, E. Klaghofer and W. Zhang (eds.) *IAHS Publication* No. 249: *Modelling Soil Erosion. Sediment Transport and Closely Related Hydrological Processes.* Oxford: International Association of Hydrology (IAHS) pp. 165–176.

van Duren, I. C., D. Boeye and A. P. Grootjans (1997). Nutrient limitations in an extant and drained poor fen: implications for restoration. *Plant Ecology* **133**, 91–100.

van Harten, H. A. J., G. M. van Dijk and H. A. M. de Kruijf (1995). *Water Quality Indicators: Overview, Method and Application.* [In Dutch; RIVM rapport 733004001.] Bilthoven, the Netherlands: National Institute of Public Health and the Environment (RIVM).

van Herwijnen, M. (1999). Spatial Decision Support for Environmental Management. Ph.D. Thesis, Vrije Universiteit, Amsterdam.

van Herwijnen, M. and P. Rietveld (1999). Spatial dimensions in multicriteria analysis. In J. C. Thill (ed.) *Spatial Multicriteria Decision Making and Analysis; A Geographic Information Sciences Approach.* Brookfield: Ashgate, pp. 77–102.

van Ierland, E. C. (1999). Environment in macroeconomic modelling. In J. C. J. M. van den Bergh (ed.) *Handbook of Environmental and Resource Economics.* Cheltenham, UK: Edward Elgar, pp. 593–609.

van Ierland, E. C. and N. Y. H. de Man (1993). *Sustainability of ecosystems: economic analysis.* [Study commissioned by the Dutch Advisory Council for Research on Nature and Environment and the Biological Council of the Royal Netherlands Academy of Arts and Sciences.] Wageningen: Department of General Economics, Wageningen Agricultural University.

Veeren, R. J. H. M. and C. M. Lorerz (2002). Integrated economic–ecological analysis and evaluation of management strategies on nutrient abatement in the Rhine basin. *Journal of Environmental Management* **66**, 361–376.

Verhallen, E. Y. (1995). *The Use of Environmental Information.* Part I: *Development of Environmental Indicators.* Bilthoven, the Netherlands: National Institute of Public Health and the Environment (RIVM) and the University of Utrecht.

Verhoeven, J. T. A. (ed.) (1992). *Fens and Bogs in the Netherlands: Vegetation, History, Nutrient Dynamics and Conservation.* Dordrecht: Kluwer.

Vreke, J. and F. R. Veeneklaas (1997). *SC-DLO Report*, No. 554: *Economische kosten-baten analyse van de Ecologische Hoofdstructuur.* [*Economic Cost–Benefit Analysis of the Ecological Main Structure.*] Wageningen, the Netherlands: Winand Staring Centre for Integrated Land, Soil and Water Research (SC-DLO).

VROM (1990). *Vierde Nota over de Ruimtelijke Ordening (Extra).* [*Fourth Document on Spatial Planning.*] The Haag: Ministerie Volksgezondheid, Ruimtelijk Ordening en Milieu (Ministry of Public Housing, Spatial Planning and Environment).

Walsh, R. G., D. M. Johnson and J. R. McKean (1992). Benefit transfer of outdoor recreation demand studies 1968–1988. *Water Resources Research* **28**, 707–713.

Walters, C. (1986). *Adaptive Management of Renewable Resources.* New York: MacMillan.

Ward, R. C. and M. Robinson (2000). *Principles of Hydrology.* London: McGraw-Hill.

Ward, R. C., J. C. Loftis and G. B. Bride (1986). The 'data-rich but information-poor' syndrome in water quality monitoring. *Environmental Management* **10**, 291–297.

Wassen, M. J. (1995). Hydrology, water chemistry and nutrient accumulation in the Biebrza fens and floodplains (Poland). *Wetlands Ecology and Management* **3**, 125–137.

Wassen, M. J., A. Barendregt, A. Paczynski, J. T. de Smidt, H. de Mars (1990). The relationship between fen vegetation gradients, groundwater flow and flooding in an undrained valley mire at Biebrza, Poland. *Journal of Ecology* **78**, 1106–1122.

Watt, K. E. F. (1968). *Ecology and Resource Management: A Quantitative Approach.* New York: McGraw-Hill.

WCED (1987). *Our Common Future.* New York: Oxford University Press for the World Commission on Environment and Development.

Weinstein, M. P., J. M. Teal, J. H. Balletto and K. A. Strait (2001). Restoration principles emerging from one of the world's largest tidal marsh restoration projects. *Wetlands Ecology and Management* **9**, 387–407.

Weitzman, M. L. (1998). The Noah's Ark problem. *Econometrica* **66**, 1279–1298.

White, R. and G. Engelen (1997). Cellular automata as the basis of integrated dynamic regional modelling. *Environment and Planning B* **24**, 235–246.

Wilen, J. E. (1985). Bioeconomics of renewable resource use. In A. V. Kneese and J. L. Sweeney (eds.) *Handbook of Natural Resource and Energy Economics*, Vol. 1. Amsterdam: North-Holland, pp. 61–124.

Wilkinson, R. (1973). *Poverty and Progress: An Ecological Model of Economic Development*, London: Methuen.

Wilson, E. O. (1998). *Consilience.* New York: Alfred Knopf.

Winsemius, P. (1986). *Guest in Your Own House. Comments on Environmental Management.* [In Dutch.] Alphen aan de Rijn, the Netherlands: Samson.

Witmer, M. C. H. (1989). *Integral Water Management at Regional Level.* Ph.D. Thesis, University of Utrecht.

Woodwell, G. M., F. T. Mackenzie, R. A. Houghton, M. Apps, E. Gorham and E. Davidson. (1998). Biotic feedbacks in the warming of the earth. *Climate Change* **40**, 495–518.

Zerbe, R. O. and D. D. Dively (1994). *Benefit–Cost Analysis: In Theory and Practice.* New York: Harper-Collins College Publishers.

Zilberman, D. and L. Lipper (1999). The economics of water use. In J. C. J. M. van den Bergh (ed.) *Handbook of Environmental and Resource Economics.* Cheltenham, UK: Edward Elgar, pp. 141–158.

Zucchetto, J. and A. M. Jansson (1985). *Resources and Society: A Systems Ecology Study of the Island of Gotland, Sweden.* New York: Springer-Verlag.

Index

acidity
  rain 30, 60, 97
  soil 21, 22–4
adaptive management 51, 62
aggregation of indicators 9, 79, 81
  environmental quality index 172–3, 175
  four-dimensional effects tables 200–2
agriculture (Vecht wetlands)
  current status 2, 97, 100, 104
Agriculture scenario
  description 108, 109–10
  future research 212
  input data 122, 131–2, 148–9
  output data 122, 123–4, 125, 137, 146–8, 153–5
  performance indicators 173, 180, 181, 188–93
  ranking 194, 195, 198, 199, 201, 202
alkalinity 30
allocation theory 44
Amsterdam 93, 102
aquifers 24, 29, 119
aquitards 24, 119
artificially created wetlands 21–2

Barendregt
benefits of wetlands 72–3, 74
benefits transfer (value transfer) 42, 69, 144
bequest values 73
Bethune polder 92, 96, 102
  development scenarios 113
biodiversity
  ecological evaluation 32–3
  landscape ecology 67–8
  as performance indicator 166–7, 169–71, 173, 175, 185–6
  and resilience 52
  wetlands 20, 21, 24–5, 28, 30–1
biological diversity 35, 36
Blue Axis 11, 100, 173, 174, 202, 204
  ecological corridors 28
bogs 7, 21, 23
bottom-up models 65, 66
brackish wetlands 20
Brundtland Commission 61

carbon cycle 5, 30
carbon dioxide 60
carrs 4, 5
CBA *see* cost–benefit analysis
cellular automata 65–6, 67
chemistry *see* water chemistry
chloride 123
'citizen responses' 42
clay soil 22, 23, 89
climate change 2
  models 60
Club of Rome 59–60
co-evolution 45
coastal wetlands *see* mangrove swamps; salt marshes
competition 25
computable general equilibrium (CGE) models 62
consilience 48–49
contingent valuation method (CVM) 41–2, 144

cost–benefit analysis (CBA) 143–6
current status *see* Vecht wetlands current status

Darcy's law 17, 118
databases 77–8
decomposition methods 43
DICE model 61
diversity *see* biodiversity
double-counting 73, 130
drainage 2, 31, 89, 97, 119, 127
drinking water 18, 96, 101–2, 119, 120
driver–pressure–state–impact–response (DPSIR) framework 49–50, 83

ecological economics 37, 43
ecological integrity 35, 36
ecological sustainability 34, 35, 36
ecology
  integration with other disciplines 47, 49–53
    see also integrated modelling
  landscape ecology 66–8
  modelling *see* vegetation response model
  wetland ecosystems 24–31
    environmental quality indicator 34
    valuation 31–4
economic model
  cost–benefit analysis 143–6
  framework 71–5, 80, 84, 86, 129–30
  input data 131–7, 148–9, 151, 210
    see also environmental quality indicator
  output data 137–43, 146–8, 153–211
  spatial equity indicator
    derivation 177–81, 188–93, 211
    evaluation 194–6, 198–204
economics
  integration with other disciplines 47, 49–53
    see also ecological economics, environmental economics, integrated modelling
  scaling issues 62, 64–6, 68–9
ecosystem health 35, 36, 46
education 34
elevated wetlands 21
emergent properties 64
employment 150
endangered species 32
energy modelling 59
environmental (business) management 37
environmental economics
  comparison with ecological economics 43
  history and development 36–7
  sustainable development 38, 40–1, 44, 45
environmental quality indicator
  derivation 34, 141–3, 163–76, 178–9, 181, 188–93, 211
  evaluation 194–7, 198, 200, 201
erosion 29
estuaries 20
European Union Habitat directive 100
eutrophication 30, 175
  as performance indicator 165, 166, 169–71, 173, 175, 184
  see also nutrient levels
evaluation, ecological 31–4
evaluation of indicators 80–1
  Vecht wetlands model
    framework 85, 86–7, 162, 163
    point evaluation 194–9
    spatial evaluation 199–204, 212
  see also performance indicators
evaporation 16
evolutionary economics 45, 53
existence values 73
externalities 37–8, 44, 50

feedback mechanisms 28–9
fens 4, 21, 23–4
floodplains (riparian wetlands) 7, 20
food webs 25–6
forested peatlands (carrs) 4, 5

game theory 43
general equilibrium models 43, 50
geographical information systems (GIS) 65, 68, 125
geomorphology of soils 22–4
global models 59–61

globalisation 39
glossary of terms 4
groundwater
  abstraction for drinking water 18, 96, 101–2, 120
  chemistry 18, 21, 121–2, 126–7
  flow 16–18, 118–19
  Vecht wetlands 97, 101–2, 117–22
  water balance 16, 29
growth theory 38, 40–1

Het Gooi 89, 93
heuristic integration 58, 63
hierarchies of scale 64–5
Horstermeer polder 93
  development scenarios 113
human influence
  ecological evaluation 33–4
  Vecht area 91–7, 164
  wetland formation/loss 5, 21–2
hydrological model
  development scenarios 110, 111–13
  framework 83, 85, 116–17
  input data 117
    chemistry 121–4, 138, 141, 148, 149, 150
    flow 117–22
  output data 117, 209
    chemistry 123, 124, 126–7
    flow 121, 122, 127
    performance indicators 163, 173, 175–6, 181–2
hydrology
  Vecht area 96–7, 101–2, 117–18, 120
  water balance 5, 15–18, 23, 29–30
  water chemistry 18–19, 20, 21, 30
  wetlands 5, 19–24, 29–30, 31

ICHORS model *see* vegetation response model
IJsselmeer 122
IMAGE model 61
indicators *see* performance indicators
input–output models 43, 50, 61–2
integrated modelling
  advantages and disadvantages 56–7
  comparison of techniques 57–62
  consilience 48–9
  framework 10–11, 49–53
  history 55–6
  performance indicators 75–80
    evaluation 80–1
  scaling issues 62–9, 80–1
  and valuation 69–75
  Vecht wetlands *see* Vecht wetlands, integrated model
isohypses 17

lakes
  Naardermeer 11, 90, 100
  reclamation 92
  recreation 102, 113
land reallotment 103–4
land use
  cellular automata models 66
  landscape ecology 66–8
  Vecht wetlands
    past 91–6
    present 13, 89–90
    scenarios in model 109, 111, 112
lexicographic preferences 43, 47
Limits to Growth model 59–60
Loosdrecht 101, 102
Lucas critique

Maartensdijk 101
macroeconomics 38, 40–1, 44, 61
mangrove swamps 29, 31
marshes 4, 19–20, 29
meta-analysis 42
Meuse 7
microeconomics 37, 64, 68
MODFLOW model 118
monetary valuation 11, 41
  and integrated modelling 69–75
moors 4
multicriteria evaluation *see* evaluation of indicators

Naardermeer 11, 90, 100
National Water Management Plan (Netherlands) 100
nature (Vecht wetlands)
    current status 89–90, 97, 103
Nature scenario
  description 108, 110–12
  input data 122, 132–4
  output data 122, 123–5, 137–9, 147–8, 156–7
  performance indicators 173, 176, 179, 180, 181, 188–93

ranking 194–5, 196, 199, 201–6, 211
Natuurmonumenten 88
net present value (NPV)
calculation 137–43, 146–7, 153–91, 210–11
derivation of spatial equity indicator 177–9, 181
ranking of scenarios 194–8, 200, 201
Netherlands
ecological planning 11, 100, 173, 174, 202, 204
land reallotment 103–4
topography 7
water balance 5, 16
water management 2–5, 7, 22
wetland formation 5–7, 19, 21–2
nitrogen 132, 138, 141, 149, 150
'no observed effect concentrations' 78
non-forested peatlands *see* bogs; fens
non-resilience (performance indicator) 167–71, 173, 175, 187
North-Holland province
land reallotment planning 100–2
NPV *see* net present value
nutrient levels (Vecht model) 123–4
eutrophication as performance indicator 165, 166, 169–71, 173, 175, 184
nitrogen 132, 138, 141, 149, 150
phosphate
balance 124, 132, 138, 141, 149, 150
removal plants 111, 122, 134

Oostvaardersplassen 144
opportunity costs 33, 72, 133, 134, 137
optimal control theory 43
oxygen 23, 30

peat
accumulation as performance indicator 165–6, 169, 173, 175, 184
formation 5–7, 21, 23
mineralisation 2
types of peatland 4, 5
utilisation 91–2
performance indicators
in integrated modelling 75–80
evaluation 80–1
Vecht wetlands model 211–12
environmental quality 141–3, 163–76, 178–9, 181, 188–93, 211
evaluation 85–7, 162, 163, 194–204, 211
net present value 137–43, 146–7
spatial equity 177–81, 188–93, 211
pH
acid rain model 60
effect on wetland formation 22, 23–4, 30
philantropic values 73
phosphate
balance 124, 132, 138, 141, 149, 150
removal plants 111, 122, 134
physics 52
piezometric head 17
pioneer phase 27
planning *see* policy
point evaluation of indicators 194–9
polders 107, 114–15
water management 22
see also Bethune *and* Horstermeer polders
policy
and integrated modelling 9, 60, 76–77
Vecht wetlands 88, 99–105
policy optimisation models 60–1
pollution
modelling 60
removal 30
Vecht wetlands 94, 97, 102–3
populations
fluctuation in 25
species valuation 32
'post-normal science' 62
precautionary principle 46
precipitation
acid rain 30, 60, 97
chemistry 18, 121–2
water balance 16, 21, 119, 120
pressure–state–impact–response (PSIR) 49–50
productivity of wetlands 20, 21, 24–5, 28, 30–1

RAINS model 60
rainwater *see* precipitation

Ramsar Convention on Wetlands
definitions 3, 5, 6
Vecht area designation 5, 11
recreation (Vecht wetlands)
current status 90, 94, 97–9
Recreation scenario
description 108, 112–13
input data 134–7, 151
output data 139–48, 159
performance indicators 141–3, 173, 176, 179–81, 188–93
ranking 194–9, 201, 202, 203–6, 211
'red lists' 32
Reference scenario (Vecht wetlands) 109
performance indicators 143, 173, 177, 179–81, 188–93
ranking 194, 195, 198, 199, 201, 202
reproduction strategies 26–7
resilience 34, 51–2, 59
as performance indicator 167–71, 173, 175, 187
resource economics 36, 37, 44
Rhine 7
riparian wetlands 7, 20
run-off 16, 20, 29–30, 132, 150

saline wetlands 19–20, 29, 31
salt marshes 19–20, 29
sandy soil 22, 89, 120
scale *see* spatial relationships
scatter diagrams 196–8
Scheldt 7
sedimentation
mangrove swamps 29
riparian wetlands 20
soil
erosion and deposition 20, 29
geomorphology 22–4
groundwater flow 16–17
Vecht wetlands 89, 120
spatial equity indicator
derivation 177–81, 188–93, 211
ranking of scenarios 194–6, 198–204
spatial evaluation of indicators 199–204, 212
spatial relationships
ecological 28, 33, 62
economic 39–40, 62, 64–6, 68–9
in integrated models 10, 62–9, 80–1, 87, 116–17, 125
spatial–ecological model *see* vegetation response model
spatial–economic model *see* economic model
specialisation 26–7
Staatsbosbeheer 88
stakeholders 3, 8–9
subsidence 2
succession 27, 28
Vecht wetlands
measure of ecosystem structure 164, 166–7
measure of resilience 167–8
sulphur 30
surface water
chemistry 20, 30, 121–4, 126–7
pollution 94, 97, 102–3
water balance 16, 29, 121
sustainable development 38, 40–1, 44, 45
swamps
definition 4
mangrove 29, 31
systems approach to modelling 58–9

TARGETS model 61
taxes 38–9
top-down models 65, 66
tourism *see* recreation
tradable permits 38
transport 40
travel cost method 41, 144
turf ponds 92

Utrecht province
land reallotment 104
planning 100–2
Utricularia vulgaris 125

valuation studies 11, 41–3
ecological evaluation 31–4
and integrated modelling 69–75
Recreation scenario 144–6
value functions 78
value transfer *see* benefits transfer
Vecht river 90, 97, 98
Vecht wetlands
agriculture *see* Agriculture scenario
current status
agriculture 2, 97, 100, 104
nature 89–90, 97, 103
recreation 90, 94, 97–9
economic activity 98–9
general description 11–13, 89–90, 107, 114–15, 120

history 90–6, 118
housing 97, 99, 104
hydrology 96–7, 101–2, 117–18, 120
industry 94, 99
integrated model
  development scenarios 85–7, 106–14
  framework 12, 13, 83–7, 116–17, 207–8
  future improvements 212–13
  indicators *see* performance indicators
  modules *see* economic model; hydrological model; vegetation response model
  outcome *see* evaluation of indicators
land use 13, 89–90, 91–6
nature *see* Nature scenario
policy 88, 99–105
pollution 94, 97, 102–3
Ramsar designation 5, 11
recreation *see* Recreation scenario
vegetation response model (ICHORS)
  environmental quality indicator
    derivation 34, 163–76, 178–9, 181, 188–93, 211
    evaluation 194–7, 198, 200, 201
  framework 84, 85, 86, 116–17
  input data 117, 123–5
    see also hydrological model, output data
  output data 125, 126, 127, 209

Wadden Sea 7
water balance 5, 15–18, 23, 29–30
water chemistry 18–19, 20, 21, 30
  eutrophication as performance indicator 165, 166, 169–71, 173, 175, 184
  modelling 121–4, 126–7
  nitrogen 132, 138, 141, 150
  phosphate
    balance 124, 132, 138, 141, 150
    removal plants 111, 122, 134
  pollution 30, 94, 97, 102–3
water management
  artificially created wetlands 21–2
  Netherlands 2–5, 7, 22
waterlogging 5, 23
welfare economics 36
wetlands
  artificially created 21–2
  chemistry 30
  definitions 3
  ecology 24–29
  formation 5–7, 19
  hydrology 5, 19–24, 29–31
  productivity 20, 21, 24–5, 28, 30–1
  research methods 8–10
  threats to 1, 2, 31
  valuation 31–4, 70–5
willingness-to-pay (WTP) 144–6
World models 59–61